W0042746

INTERNATIONAL CENTRE FOR MECHANICAL SCIENCES

COURSES AND LECTURES No. 201

Ro. man. sy, '73

FIRST CISM - IFToMM SYMPOSIUM
5 - 8 September 1973

ON THEORY AND PRACTICE
OF
ROBOTS AND MANIPULATORS

VOLUME I

UDINE 1974

SPRINGER-VERLAG WIEN GMBH

This work is subject to copyright.

All rights are reserved,

whether the whole or part of the material is concerned

specifically those of translation, reprinting, re-use of illustrations,

broadcasting, reproduction by photocopying machine

or similar means, and storage in data banks.

© 1972 by Springer-Verlag Wien

Originally published by Springer-Verlag Wien New York in 1972

ISBN 978-3-211-81252-5   ISBN 978-3-7091-2993-7 (eBook)

DOI 10.1007/978-3-7091-2993-7

ORGANIZING COMMITTEE

Chairman:
Prof. A.E. KOBRINSKII
Academy of Sciences of the USSR, ul. Griboedova 4, Moscow (Centre), 101000 (USSR)

Vice-Chairman:
Prof. L. SOBRERO
CISM, Palazzo del Torso, Piazza Garibaldi, Udine (Italy)

Members:
Acad. I.I. ARTOBOLEVSKII
Academy of Sciences of the USSR, ul. Griboedova 4, Moscow (Centre), 101000 (USSR)

Prof. G. BIANCHI
Politecnico, Piazza L. da Vinci 32, 20133 Milano (Italy)

Prof. I. KATO
Waseda University, Faculty of Science and Engineering, Nishiookubo, Shinjuku-ku - Tokyo (Japan)

Prof. M.S. KONSTANTINOV
IFToMM Secretariat - P.O. Box 431 - Sofia C. (Bulgaria)

Prof. A. MORECKI
Technical University, Nowowiejska Street 22-24, Room 206 - Warsaw (Poland)

Prof. A. ROMITI
Politecnico, Corso Duca degli Abruzzi 24, 10129 Torino (Italy).

Prof. B. ROTH
Stanford University, Dept. of Mechanical Eng. - Stanford - California 94305 (U.S.A.)

Prof. M.W. THRING
University of London, Queen Mary College, Dept. of Mechanical Eng., Mile End Road - London E.1. (U.K.)

Dr. VUKOBRATOVIĆ
Institut "Mihajlo Pupin", Volgina 15, 11000 Beograd (Jugoslavia)

Prof. H.J. WARNECKE
Institute for Production and Automation, University of Stuttgart, P.O. Box 951 - Stuttgart (F.R.G.)

Assoc. Prof. D.E. WHITNEY
Massachusetts Institute of Technology, Dept. of Mechanical Eng. Room 1-110 - Cambridge, Massachusetts 02139 (U.S.A.)

Secretary :
Mrs. A. Bertozzi
CISM, Piazza Garibaldi 11 - 33100 Udine (Italy) - Telephone 0432-64989 - 22523

PREFACE

Trying to sum up the results of our Symposium, I recall some facts, pertaining to its organization and some events of two years ago, in particular.

Then, in Cupari, Yugoslavia — a picturesque corner at the Mediterranean — in the course of the regular IV Congress of our Federation — a decision was taken to call this Symposium. At the same time we met with professor Sobrero, our host of to-day, the chief secretary of CISM — the organization that led with us the enormous work whose fruits we reap at present.

This Symposium was the first of its kind for CISM, as well as for IFToMM. As I recall it now, already then, in Cupari, the initiative group of CISM — IFToMM had no doubts about the large theoretical and practical importance of such Symposium. Already then it was quite clear, that the technique of robots will be in the nearest future of great applied and practical importance. At the same time, the volume of new problems, that the technique of robots has already and will put in the near future before the theoreticians-specialists in the field of applied mechanics, biological mechanics, mathematics, theory of control — was already then intuitively felt by all of us.

To-day I can mention with pleasure that our choice was absolutely correct. The theory and design problems of manipulators and robots — a whole class of machines of a new type — are just those problems to which may be widely and efficiently applied theoretical and experimental methods in which so rich is the theory of machines, the biological mechanics and other branches of mechanics including the mechanics of robot and manipulator movement control.

Our Symposium has quite obviously shown that these methods have already worked their way into the field of robot techniques. This is the first and, as it seems to me, a very important conclusion, that we may draw to-day.

But this conclusion is not the only one that is, according to my opinion, very important. To-day many well known scientists and specialists working in different fields of science have met in this hall. How fruitful these meetings are — has become obvious from the first years of the formation of the science of cybernetics. But in order to realize this possibility these meetings must not be of an abstract but of a concrete character. They must be devoted to the study of objects having equally "burning" interest for different specialists. It is necessary that all attending such meetings, appear simultaneously as research workers in the field, they are interested in, as well as consulting workers for persons working in adjoining fields of science.

From this point of view it may happen, that these objects are not so numerous and among those that it is possible to enumerate, there are quite few ones that have a great applied importance and wide outlooks for their being used in future.

Objects under our study — robots and manipulators — are just such objects. And for all of us — specialists working in adjoining fields of science — they are of "burning" interest. We listened to and discussed with great interest and advantage many reports devoted to the structure, kinematics and dynamics of robots and manipulators, as well as to biomechanics and to problems of such systems control and to the creation of man-made (artificial) intellect, that may be used in these systems.

That is why it is possible to draw the conclusion that this Symposium was very important not only for the development of problems connected with the theory and principles of robot and manipulator design, but was also very useful and fruitful for all participants from the point of view of interchange of experience, as well as interchange of new results and ideas.

These conclusions, arrived at as a result of our work, show, that it is extremely expedient to maintain close contacts among specialists, concentrating their efforts on problems of robot techniques. I am quite certain that ROMANSY-73 will be followed by other symposiums. I think we have found true forms of holding them. We determined precisely and at the same time widely enough the subjects to be discussed. In this respect, I think, in future there will be no need to introduce many changes and the organization committee of the next Symposium will have less difficulties than we had, though it will have a lot of work to do.

In conclusion I should like to thank on behalf of the Soviet Union delegation as well on behalf of our federation our kind hosts — the CISM organization and personally professors Sobrero and Bianchi for the splendid organization of the Symposium and for our cordial and affectionate reception.

I.I. Artobolevskii
President of IFToMM

The "First CISM-IFToMM Symposium on Theory and Practice of Robots and Manipulators" (Ro.man.sy'73) was programmed at a meeting held at Kupari (Jugoslavia) during the month of September 1971 between Professors Artobolevskii (Moscow), Kobrinskii (Moscow), Konstantinov (Sofia), Roth (Stanford) and Bianchi (Milan) for IFToMM, and Professor Sobrero and Mrs. Bertozzi for CISM. It was then outlined in its fundamental features at Nieborów (Poland) in May 1972 with the presence of Professors Kobrinskii (Chairman of the Organizing Committee), Artobolevskii, Konstantinov, Roth, Thring (London), Morecki (Warsaw), Vukobratović (Belgrade), Kato (Tokyo) and Bianchi for IFToMM, and Professor Sobrero and Mrs. Bertozzi for CISM. Furthermore, it was defined in detail at a meeting held for this purpose at Split in April 1973, with the attendance of the above mentioned Professors, as well as Professors Bazjanac and Jelovac (Zagreb), of Professor Warnecke (Stuttgart) and of Professor Romiti (Turin).

We can say that a scientific congress has rarely been organized with greater care: at the Split meeting the work of the different lecturers was examined, that which was not satisfactory was discarded, it was decided that the chosen works should be published as pre-prints (45 altogether), to be distributed amongst the participants at the symposium and finally published in their final form, Springer-Verlag of Vienna-New York would be commissioned with the sale of same. The success of the Symposium has been extremely encouraging, the participation of scientists from all over the world exceeded all expectations.

At the end of the Symposium it was decided that the previous organizing committee should stand, and the chairmanship passed to Prof. Roth. A new Symposium is to be held in Poland in 1976. Previous to this Symposium, on account of the valuable experience gained, two preliminary meetings will be held, the first of which will be held in Warsaw in June 1974.

Udine, March 1974.

L. Sobrero
Secretary General of CISM

OPENING LECTURES

"THE STATE OF THE ART IN THE FIELD
OF ROBOTS AND MANIPULATORS"

Mr. Chairman!
Ladies and gentlemen!

In 1920 the Czech writer Karel Čapek wrote the play "RUR." He called it "A Collective Drama." Side by side with ordinary people you see industrially manifactured mechanical "pseudo-men" acting in it.

Today there is no need to render the contents of this work which is well known to the world public at large.

For half a century the word "robot" appeared on the pages of science fiction stories and novels later to be used in scientific and technical articles. This name was given to machines of a new type, which first slowly, but then faster and faster began to be introduced into different fields of technology. The early machines of this type appeared about a quarter of a century ago. They were called manipulators or mechanical hands. These devices enabled man to handle radioactive materials without direct contact with them. As the uses of atomic energy broadened the assortment of mechanical hand increased too. The movements of these devices were controlled by operators placed in a safe zone adequately protected against radioactive radiation.

The World Ocean accounts for more than 70 per cent of the Earth's surface. Its wealth is boundless and has always attracted the attention of man.

However, the world record depth set by a skin diver without a breathing apparatus (the American Croft) is 73 metres.

Divers in special suits can go down to depths of about 100 metres and work there for several hours.

Now the larger area of the World Ocean is characterised by depths ranging from two to six kilometres. To get to the wealth hoarded there and to recover it you need deepsea diving apparatuses outfitted with mechanical hands. Starting with Professor Picard's **Trieste** such apparatus have been built and used in several countries for a period of many years.

Space exploration and research offer a wide field of application for robots. Already now we are observing the motions of thousands of man-made celestial bodies outfitted with a wide range of equipment which needs maintenance and servicing. To this end there is no method other than the use of specialised machines capable of transmitting the movements and efforts of man over cosmic distances or capable of reproducing these movements and efforts automatically.

In the last ten years robots have appeared in industrial production, i.e. in the field of technology which, though characterised by a relatively high level of automation, still employs the manual labour of tens of millions of people. It is worth noting that this labour often requires high energy expenditure, it is tiresome, monotonous and fraught with occupational diseases. The fields in which robots can be applied in industry are virtually boundless. That is why we can positively assert that the "boom" in the sphere of robot equipment which we observe today is not mere tribute to vogue.

Robotization is a stable line in mechanization and automation of human labour, a most important direction in scientific and technological progress for many years to come.

I have mentioned four different fields of application of machines of a new type, namely atomic energy, exploitation of the wealth of the World Ocean, space exploration and modern industrial production. What features do these machines have in common, though they are designed for work in such seemingly different spheres?

Wherein lie the "humanoid" features of these machines? Really you cannot seriously speak of "humanoid" (or "anthropoid") appearance of the modern industrial robot!

Then why has the term which implies something "humanoid" stuck to them so fast?

It appears that we can give the following answer to this question. These machines are "humanoid" functionally, and not in appearance. They are capable of executing certain motions and actions of human character to a certain degree of perfection. Having seen how they work you cannot help calling them robots.

The human hand is a universal implement of labour. It is capable of carrying out an infinite number of the most varied movements - simple and complex, quick and smooth, sweeping and minute. These motions are controlled by the human brain. Though very much has been written about the principles of design of this device, very little is still known about it.

Robots and manipulators of different kinds also work with hands, but mechanical hands, of course. The range of their movements and functional potentialities are, naturally, inferior to those of their human counterparts. However, they too are capable of performing a wide variety of movements.

What are the methods for the reproduction of these movements by mechanical hands? There are two such methods. The first - both in time and so far in scope of application - is the method of operator control, i.e. when the movements of the mechanical hands are controlled by the operator. The second which has been gaining considerable scope in the recent years (I have in mind industrial robots) is the method of programmator or computer control.

Manually and automatically controlled robots and manipulators today form a large and highly promising class of machines. Questions of the theory and application of these machines form the contents of the work of the present symposium. I shall try to give a brief review of the essence of these questions.

Nature developed the design of the human hand in the course of many millenniums.

Structurally the hands of all people are the same, the relative sizes of the elements varying within very narrow limits. Inasfar as the kinematic, dynamic and precision characteristics of the human hand are concerned, we, naturally, realise that they differ widely in the case of the violinist, lumberjack, engraver and weight lifter. And yet they are always within the scope of human potential.

As to mechanical hands the matter is totally different. In the various kinds of robots and manipulators the structure and relative sizes of the elements are different. The kinematic dynamic and other characteristics are different too.

How should you compare the different robot designs? How should you select a structure, size and type of joints to produce better mechanical hands? How should you calculate their kinematic and dynamic characteristics? These are big questions and several papers at the present symposium are devoted to them. When solved these questions will enable us to calculate the design of the existing machines and develop optimal designs of future robots and manipulators with the help of the quickest and most economical method.

To solve these questions it is necessary to resort to methods of the theory of machines and mechanisms - already classical - and to new methods which have been developed and are being developed by specialists in the field.

The two different methods of control of artificial hands — manual and automatic — used in manipulators and robots require the development of methods of analysis and synthesis of the corresponding movement control systems.

Such systems as the "operator-robot," "operator-manipulator" or "patient-prosthesis" are biotechnical systems whose human and technical components are intimately connected in the process of work. In these systems it is the operator that lays down the entire programme of movements and actions of the artificial hands. At first glance it may appear that the development of such movement control systems should not give rise to any special difficulties.

However, in reality everything is absolutely different. The speed and accuracy with which the movements and effort are transmitted in such systems depend on a large number of different factors. I do not intend to enumerate or analyse them. I shall only say that the quality and efficiency of such hand control systems are determined by the degree of accuracy and convenience achieved in the adjustment of the two parts that make up the system.

Specialists engaged in developing master-slave manipulators, exoskeletons and prostheses are quite right in asserting that the elaboration of the theory and methods of analysis and synthesis of such devices has brought forth new problems of considerable importance in the sphere of biophisics and, in particular, biomechanics. Only sound knowledge of the motorial potentialities of man, of his kinematic, dynamic, energy and accuracy characteristics will make it possible to produce the so-called biomechanical conditions. These are necessary, because they enable the scientist and engineer in each given case to choose the optimal parameters for these comprehensive systems, instead of groping in the darkness. That is precisely why our symposium is devoting such attention to questions of biomechanics.

However, it should be pointed out that knowledge of the biomechanical conditions alone will not provide the solution to the whole of the problem. It is here that specialists in the control theory will be required to make their contribution.

Manipulator and robot-type machines are characterized by three specific features:

1) Large number of degrees of freedom inherent in these machines. The operator has to effect simultaneous control of movements in keeping with 6-8-10

and sometimes even dozens of independent coordinates.

2) In many cases the imperative condition is laid down that the operator should "feel" the magnitude of the resistence offered to the mechanical hands the operator is manipulating. As they say, these control systems should have the quality of reversibility.

3) The third specific feature which complicates the preceding two even more consists in the fact the movements and efforts from the operator have to be transmitted to the machine he is operating over various distances: in the first case, over distances of the order of several metres or several dozen metres (as in work in hot chambers); in the second case, over distances of the order of several hundred or several thousand metres (in deep-sea works) and, in the third case, over distances of the order of several thousand kilometres or several hundred thousand kilometres (in space vehicles).

Yes, indeed specialists in control are confronted with many problems in the theory of robotry. That is why they are taking such an active part in the work of our symposium.

I do not intend to deal with questions bearing on the theory, principles of construction and methods of analysis and synthesis of information channels for effecting control of robot movements. The reason for this is not that we fail to realise or that we underestimate their importance. The real reason is that the scope of problems to be studied is so wide and its nature is so specific that it merits separate detailed consideration. Besides, these problems are within the competence of other national and international scientific organizations, such as IFAC and IFIP which are already in work on the theory of robotry.

In speaking about the working elements of robots I mentioned mechanical hands only. Now if we intend to speak seriously about the broad use of robots, if we bear in mind that they should not be confined to one place alone, because this considerably narrows their potentialities, that ideas have been advanced about robots coming to our homes in the future, we will see that the questions of theory and principles of design of the lower limbs, and not only of the upper limbs, are of major importance. Until recently these questions were of topical character only in the field of prosthesis, perhaps. Today their practical importance has grown immeasurably.

Our civilisation has developed in a way that the interior design of the structures, buildings and homes is adapted chiefly for stepping (walking) motion, whereas the communications between them are designed for rolling motion. The staircases, thresholds, narrow corridors with turns and doors in the long sides, shops and offices furnished with equipment and furniture are absolutely unsuited for rolling motion. The question being: should a robot walk or roll into a shop or home, the specialists working on the theory and principles of walking machine design have highly promising prospects before them. Though it is sad, the problems of lower limb prosthesis and patient equipment design will, apparently be of interest to humanity for a long time to come. That is why our symposium has devoted a special sitting to questions of analysis and synthesis of walking machines and to questions of control of such machines.

How do matters stand with robots that are capable of acting on their own, of moving and working without man's direct participation, automatically performing movements and applying measured effort? In general, what are the problems that are of interest to us in connection with machines which fit more fully our idea of a "real" robot?

As to the industrial robots, they still have no feedback with environment. In this sense they can be regarded as the first generation of such machines. For several years now many countries have been trying to develop robots of the next generations which would be provided with artificial senses, which would be capable of interacting with environment, which would be characterised by a certain degree of self-sufficiency in an environment which is not completely organized.

It appears that the development of the theory and principles of design of such machines includes two chief lines, namely:

1) Development of artificial senses, sight for instance, sense of touch and corresponding sensibilization. This line, apparently is close to the problem of devising information channels. It will not be considered at our symposium.

2) The second line is connected with the problem of so-called "artificial intellect." The conception of "artificial intellect" is still defined very ambiguously. "Artificial intellect" is mentioned in connection with automatic writing of poetry or playing chess. Out of the vast scope of questions which are covered by the conception "artificial intellect" we are interested only in question connected with the formation of robot movements, with the development of algorithms for control of movements and force magnitude, with the establishment and distribution of robot control functions between the computer and the man in supervisor control systems. Specialists working in the field of the control theory and applied mathematics will cover these questions in the papers to be presented to the

symposium.

The idea about the scope of the work of our symposium would not be complete, unless we mention several papers which are chiefly devoted to questions of robot and manipulator applications. The authors of these papers are engaged in the operation of such machines and in the development of the scientific foundations for the technique of robot application. They thus pose before us a series of new theoretical problems.

As you see, we have a lot of interesting work to do. The question may arise: Are not we afraid of such a wide variety of problems confronting us? Will the theorists and practical workers be able to find a common language, the specialists in theory of machines and theory of control, the biomechanics and applied mathematicians? Won't our sessions turn into another Babel?

No, we are not afraid of this! The Biblical events took place a very long time ago and a lot has changed since then. The engineers and scientists are already accustomed to the idea that most of the modern problems are of a complex multi-disciplinary character, they call for contributions from different specialists and a system approach. This is an inevitability conditioned by the requirements of the times and the broader the scope of such interaction and the more active its character the better it is for our undertaking.

The distance separating Karel Čapek from cybernetics was covered in a quarter of a century. The next twenty five year period built the foundation for what seemed pure fantasy. We maintain that the next quarter of a century will be a period of quantitative accumulation and a qualitative leap in the field of robotry.

People were always eager and have earned a right to do skilled and interesting work that presented no hazard to their life or health, work that would give them maximum satisfaction and maximum benefit to society. Robots are giving us the possibility to realize this right in practice. I am confident that in another twenty five years people will be looking at our present day robots and manipulators as we are now looking at the early automobiles and the Wright brothers' plane — with respect and with a smile at the same time.

Concluding, I would like to say that when you look at a working machine, regardless of whether it is a car, aircraft or robot, you should pay tribute to the engineers and designers that developed and produced it. It is a fact that develop a modern machine intuition, experienced and professional skills are not enough. Science and theory play a tremendous role in modern technology, in particular in robotry. That is why I think it is an honour and a privilege to take part in the work of our symposium.

Thank you for your attention.

A.E. Kobrinskii
Academy of Sciences
of the USSR

ROBOTS AND MANIPULATORS

Man has always dreamed of a mechanical slave of giant strength which would carry out his wishes at a word of command. The Genie of Alladdins Lamp, the Golem of Prague, Mary Shelley's Frankenstein and most recently Rossum's Universal Robots of Karel Čapek all express this idea but also stress the danger if such powerful machines fall into evil hands. Modern Science fiction has gone much further with this idea and the dangers of machines running amok.

As soon as one makes a proper engineering study of the possibilities of building powerful robots one realizes that machines are no more than the human designer and constructor makes of them. If they are built for humane purposes and the designer foresees all the possible ways they can go wrong then he can design safety measures to prevent those dangers , for example he can build in a device so that a human voice shouting 'stop' can switch off all power in the robot so that it freezes in position. Like all machines they merely amplify the powers of the user and can be used for good or evil purposes.

One can define the field of this conference as all machines that have limbs similar to those of a human or an animal, particularly arms, hands and legs. I think it is useful to make a family tree for such machines, which are dealt with at the conference

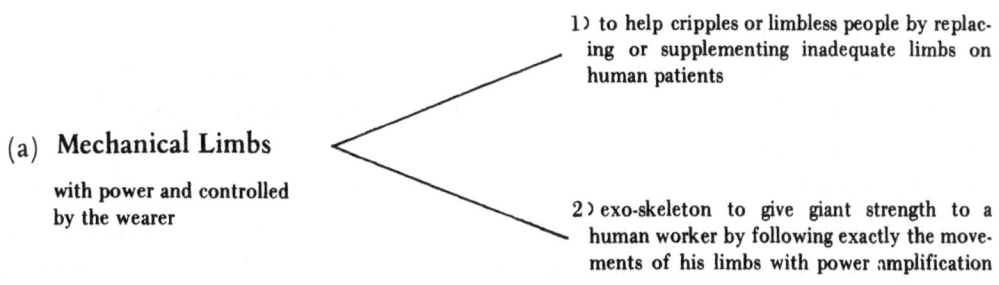

(a) **Mechanical Limbs**

with power and controlled
by the wearer

1) to help cripples or limbless people by replacing or supplementing inadequate limbs on human patients

2) exo-skeleton to give giant strength to a human worker by following exactly the movements of his limbs with power amplification

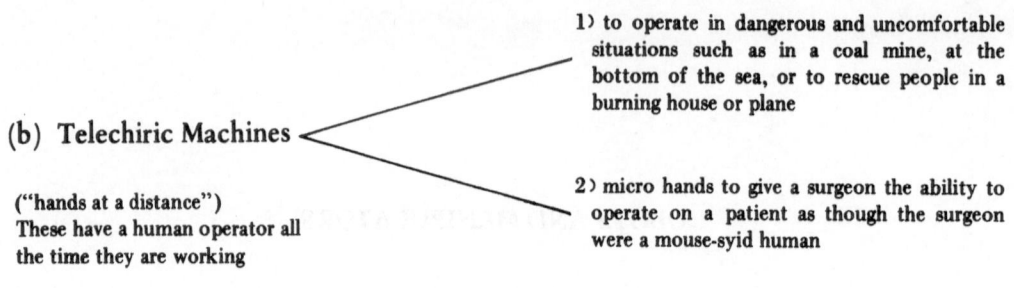

(b) Telechiric Machines

1) to operate in dangerous and uncomfortable situations such as in a coal mine, at the bottom of the sea, or to rescue people in a burning house or plane

("hands at a distance")
These have a human operator all
the time they are working

2) micro hands to give a surgeon the ability to operate on a patient as though the surgeon were a mouse-syid human

(c) **Robots** Machines which have been pre-instructed by a human to carry out a series of complex manipulative operations. They are small computers with arms and legs, and are trainable morons with great strength, reliability, tirelessness, docility and memory for exact movements.

Nearly all the robots that are in regular use are 'senseless' that is to say they repeat the movements that they have been instructed to do even if the object they have to handle is not in position. Nevertheless they are already being used in many tasks of a highly repetitive character in industrial production on assembly lines or loading and unloading furnaces. They can work without rests or breaks 24 hours a day in very uncomfortable positions or positions unaccessible to ordinary hands.

Most of the research on robots is to give them senses such as sight and touch and then to give them sufficient computer facility so that the human instructor can tell them to vary their movements according to their sense observations such as the object to be handled being in a variable place, upside down or faulty.

It is my opinion that Robots can never be developed as in Čapek's play RUR to think for themselves and initiate their own motivation but in any case even if we could, we should be very foolish to design such facilities into them. Nor is there any sense in designing them to look like human beings; they should be designed to make the best engineering use of the materials and mechanisms available. They can therefore be developed to act as true slaves of humans to relieve them of drudgery (all repetitive unthinking tasks in the home and factory and office) and of all dangerous and uncomfortable work but must be very careful not to allow them to damage human life in any way e.g. by causing unemployment.

M.W. Thring
University of London

LIST OF PAPERS

1. **Walking Machines**
   A.P. BESSONOV - N.V. UMNOV
   "The analysis of gaits in six-legged vehicles according to their static stability"

   I. KATO and Group of Bio-Engineering (Waseda University)
   "Information-Power machine with senses and limbs - WABOT 1"

   R.B. McGHEE - D.E. ORIN
   "An interactive computer-control system for a quadruped robot"

   T. YAMASHITA - H. YAMADA
   "A study on stability of bipedal locomotion"

2. **Kinematics and Dynamics**
   I.I. ARTOBOLEVSKII - A.G. OVAKIMOV
   "A generalized method for solving a group of problems referring to the dynamics of manipulators with an electromechanical servo drive"

   M.S. KONSTANTINOV - Z.I. ZANKOV
   "A kinematical algorithm and dynamical point mass simulation applied in robots and manipulators"

   B. ROTH - J. RASTEGAR - V. SCHEINMAN
   "On the design of computer controlled manipulators"

3. **Biomechanics of Motion**
   A. CAPOZZO - T. LEO
   "Biomechanics of walking upstairs"

   V.S. GURFINKEL - S.V. FOMIN
   "Biomechanical principles of constructing artificial walking systems"

   A. MORECKI - K. FIDELUS
   "On analysis and synthesis of distribution of drives in live manipulators and pedipulators"

V. PAESLACK - H. ROESLER
"Medical manipulators - general considerations on the design criteria of
manipulative technical aids for quadriplegics"

Y. UMETANI - S. HIROSE
"Biomechanical study of serpentine locomotion "

I.B. VINOGRADOV
"Kinematic features of human arm in "operator-manipulator" system"

4. Man - Machine Systems
A. FREEDY - F. HULL - G. WELTMAN - J. LYMAN
"The application of sensory information and multifunction learning to
autonomous manipulator control"

A.E. KOBRINSKII et al.
"Analysis of quality of manipulator manual control"

A. MORECKI et al.
"Manipulators for supporting and substituting lost functions of human extremities"

E.P. POPOV - N.A. LAKOTA
"Design of master-slave manipulators: biotechnical aspects"

Y. Yu. SHISHMAREV et al.
"Normalization of walking on prothesis with an external power source"

R. TOMOVIĆ
"Logical control of robots"

G WELTMAN - A. FREEDY
"Man-machine interactions in intelligent robotic systems"

J. VERTUT
"Contribution to analyse manipulator morphology coverage and dexterity"

5. Artificial Intelligence
M.V. ARISTOVA - M.B. IGNATIEV
"The concept of the structure of highest levels control in robot manipulators"

C.R. FLATAU
"The synthesis and scaling of advanced manipulator systems"

7. Control of Motion
V.P. DOROKHOV,
"Master-slave remote-control manipulator"

M. GAVRILOVIĆ - M. MARIĆ
"New developments in synergic rate control of manipulators"

V.S. KULESHOV - V.N. SHVEDOV
"Robot and manipulator slave from control viewpoint"

M.L. MOE
"Kinematics and rate control of the Rancho arm"

J.L. NEVINS - D.E. WHITNEY
"The force vector assembler concept"

K. SATO - K. OKAMOTO - K. TAKASE - H. INOUE
"Articulated manipulators and integration of control software"

T.B. SHERIDAN
"On modeling performance of open-loop mechanisms"

M. VUKOBRATOVIĆ
"Dynamics and control of antropomorphic active mechanisms"

J.F. YOUNG
"Robot vision as a factor in the control of motion"

Survey Papers
H.J. WARNECKE
"Industrial robots - Problems of design and application"

LIST OF AUTHORS

A.D. Alexander — NASA Advanced Concepts and Missions Division, California, USA.

M.V. Aristova — Leningrad Institute of Aviation, Leningrad, USSR

M.V. Artiushenko — Institute of Cibernetics, Kiev, USSR .

I.I. Artobolevskii — Academy of Sciences, Moscow, USSR .

H. Borowski — Warsaw Technical University, Warsaw, Poland

A.P. Bessonov — Institute for the Study of Machines, Moscow, USSR

A.I. Bogomolov — The Central Research Institute for Prosthetics, Moscow, USSR .

Z. Busko — Warsaw Technical University, Warsaw, Poland

A. Capozzo — University of Roma, Roma, Italy

S. Deutsch — National Aeronautics and Space Administration, Washington

V.P. Dorokhov — State Committee for Atomic Energy, Moscow, USSR

K. Fidelus — Academy of Physical Education, Warsaw, Poland

C.R. Flatau — Shorearm, N.Y., USA .

S.V. Fomin — Academy of Sciences, Moscow, USSR

A. Freedy — University of California, California, USA

M. Gavrilović — Mihailo Pupin Institute, Belgrade, Yugoslavia

A. Gill — University of Tel Aviv, Telaviv, Israel

V.S. Gurfinkel — Academy of Sciences, Moscow, USSR

E. Heer — Jet Propulsion Laboratory, Pasadena, California, USA

S. Hirose — Tokyo Institute of Technology, Tokyo, Japan

F. Hull — University of California, Los Angeles, USA

M.B. Ignatiev — Leningrad Institute of Aviation, Leningrad, USSR

H. Inoue — Electrotechnical Laboratory, Chiyoda-ku, Tokyo, Japan

I. Kato — University of Waseda, Tokyo, Japan

H. Kobayashi — University of Waseda, Tokyo, Japan

A.E. Kobrinskii — Institute for the Study of Machines, Moscow, USSR

A.I. Korotkov — The Central Research Institute for Prosthetics, Moscow, USSR

M.S. Konstantinov — Mechanical and Electrotechnical Institute, Sofia, Bulgaria

E. Kotwicki — Warsaw Technical University, Warsaw, Poland

D.A. Kugath — General Electric Company, Philadelphia, Pennsylvania, USA

A.I. Kukhtenko	Institute of Cibernetics, Kiev, USSR
V.S. Kuleshov	Academy of Sciences, Moscow, USSR
N.A. Lakota	Academy of Sciences, Moscow, USSR
T. Leo	University of Roma, Italy
J. Lyman	University of California, Los Angeles, USA
F. Luccio	University of Pisa, Pisa, Italy
G.I. Lukishov	State Committee for Atomic Energy, Moscow, USSR
T.B. Malone	Essex Corporation, Alexandria, Virginia, USA
M. Marić	Mihailo Pupin Institute, Belgrade, Yugoslavia
R.B. McGee	Ohio State University, Columbus, USA
Yu. V. Miloserdin	Engineering Physics Institute, Moscow, USSR
M. Moe	University of Denver, Denver, Colorado
A. Morecki	Warsaw Technical University, Warsaw, Poland
I.S. Moreinis	The Central Research Institute for Prosthetics, Moscow, USSR
E. Mühlenfeld	Institute fur Informationsverarbeitung in Technik und Biologie, der Fraunhofer-Gesellshaft, West Germany
J.L. Nevins	Draper Laboratory, Cambridge, Massachusetts, USA
S. Ohteru	University of Waseda, Tokyo, Japan
K. Okamoto	Electrotechnical Laboratory, Chiyoda-ku, Tokyo, Japan
D.E. Orin	Ohio State University, Columbus, USA
A.G. Ovakimov	Moscow Aviation Institute, Moscow, USSR
V. Paeslack	Orthopädische Klinik und Poliklinik der Universität Heidelberg, Heidelberg, Germany
R. Pasniczek	Medical Academy, Warsaw, Poland
J.R. Parks	National Physical Laboratory Teddington, England
R. Paul	University of California, Los Angeles, USA
E.P. Popov	Academy of Sciences, Moscow, USSR
A.N. Radchenko	Politechnical Institute M.I. Kalinin, Leningrad, USSR
J. Rastegar	Stanford University, Stanford, California, USA
H. Roesler	Orthopädische Klinik und Poliklinik der Universität Heidelberg, Heidelberg, Germany
B. Roth	Stanford University, Stanford, California, USA
K. Sato	Electrotechnical Laboratory, Chiyoda-ku, Tokyo, Japan
V. Scheinman	Stanford University, Stanford, California, USA
V.N. Semenov	Institute of Cibernetics, Kiev, USSR

V.I. Sergeev	Institute for the Study of Machines, Moscow, USSR
T.B. Sheridan	Massachusetts Institute of Technology, Cambridge, Massachusetts, USA
L.I. Slutskii	Institute for the Study of Machines, Moscow, USSR
V.N. Shvedov	Academy of Sciences, Moscow, USSR
V. Yu. Shishmarev	The Central Research Institute for Prosthetics, Moscow, USSR
K. Shirai	University of Waseda, Tokyo, Japan
K. Takese	Electrotechnical Laboratory, Chiyada-ku, Tokyo, Japan
K. Tempinski	Warsaw Technical University, Warsaw, Poland
M.W. Thring	University of London, London, England
R. Tomović	University of Belgrade, Belgrade, Yugoslavia
L.I. Tyves	Institute for the Study of Machines, Moscow, USSR .
A. Uchiyama	University of Waseda, Tokyo, Japan
Y. Umetani	Tokyo Institute of Technology, Tokyo, Japan
N.V. Umnov	Institute for the Study of Machines, Moscow, USSR
J. Vertut	Commissariat à l'Energie Atomique, Saclay, France
I.B. Vinogradov	Institute for the Study of Machines, Moscow, USSR
M. Vukobratović	Institute Mihailo Pupin, Belgrade, Yugoslavia
H.J. Warnecke	University of Stuttgart, Federal Republic of Germany
G. Weltman	University of California, Los Angeles, USA
D.E. Whitney	Massachusetts Institute of Technology, Cambridge, Massachusetts, USA
D.R. Wilt	General Electric Company, Philadelphia, Pennsylvania, USA
H. Yameda	Kyushu Institute of Technology, Kitakyushu, Japan
T. Yamashita	Kyushu Institute of Technology, Kitakyushu, Japan
J.F. Young	University of Aston, Birmingham, England
E.I. Yurevich	Politechnical Institute M.I. Kalinin, Leningrad, USSR
Z.I. Zankov	Mechanical and Electrotechnical Institute, Sofia, Bulgaria

1. WALKING MACHINES

7. WALKING MACHINES

THE ANALYSIS OF GAITS IN SIX-LEGGED
VEHICLES ACCORDING TO THEIR STATIC STABILITY

A.P. BESSONOV, Professor
Institute for the Study of Machines,
Moscow, USSR

N.V. UMNOV, Senior Researcher,
Institute for the Study of Machines,
Moscow, USSR

(*)

For the sake of convenience the study of walking vehicles may be subdivided into three large groups according to the number of legs attached to each vehicle, namely, biped, quadruped and six-legged. Each of these means of transport has its merits and demerits. For example, biped vehicles will probably reveal utmost manoeuvrability and land mobility and least stability, and, as a result of the latter, require a most complex control system for balance maintenance.

In any case moving two of the vehicle's legs excludes the possibility of moving in a state of continuous static stability. This mode of movement is given sufficient attention in the study of quadruped vehicles, for the advantage of such mode of walk is quite evident. Leaving aside the simplification of the control system, owing to the fact that the static balance is provided merely by an adequate choice of a legs' operation regime and a requisite pattern of their operation, one ought to bear in mind a possibility of stopping the vehicle's movement at any given moment.

However, the studies of several authors [1-2] show that the stability of statically balanced gaits in quadruped vehicles is not sufficient. The frame's centre of gravity is projected close to the boundary of the supporting polygon and the stability reserve is not great. Hence, the recent interest of scientists in multipod machines, particularly in six-legged vehicles [3-4] is fully justified.

(*) All figures quoted in the text are at the end of the lecture.

The difficulties of the study of gaits in six-legged vehicles may be ascribed mainly to their multiplicity. For example, a quadruped variety reveals only 6 different regimes of orderly leg operation and only one of them provides for the vehicles continued static stability [5]; in all these gaits the legs operate in turn and a simultaneous operation of two legs cannot be performed for in this case there may be moments when only two legs will be in touch with the earth. Statically stable gaits in six-legged vehicles, apart from a single leg operation, following a periodically alternating pattern, incorporate those gaits in which legs may operate in pairs, and even, three legs at a time (for example, the widely known "tripled" gait).

Bearing in mind that 120 is the possible number of various patterns of single leg operations, the number of simultaneously operating pairs and triples would increase this theoretically to 1030 which is too great for analysis.

However, by far not all of the 1030 gaits provide for the vehicle's continued static stability. On the one hand, it may happen that at certain moments, under unfavourable conditions, less than three legs are on the earth, and, on the other hand, in some gaits the projection of the centre of gravity may run beyond the boundary of the supporting polygon.

This study is aimed at estimating those particular gaits in six-legged vehicles which furnish continuous static stability in the course of walking.

A comparative study of walking was applied to a definite vehicle pattern (Fig. 1a) as follows: six weightless legs were fixed in such a way that 3 were attached to each side. The points of the suspended legs on one side were aligned at equal distances from each other. The authors examined the vehicle's uniform rectilinear motion; as a result the trajectories of the leg's supporting ends (fixed in a coordinate system connected with the frame) being projected onto a horizontal plane will appear to be straight lines (tracks).

The tracks of all legs are believed to be alike and the neighbouring tracks of each side are lacking gaps.

The centre of gravity of the entire vehicle coincides with the frame's centre of gravity and is situated in the centre of the line of suspension of mean legs in the vehicle's longitudinal axis.

The leg's kinematic scheme, its drive system and features of design are of no importance, and, are omitted. We take into account merely the phases of the leg's operation (support and transfer phases). The leg partakes in the vehicle's maintenance at the support phase, whereas throughout the transfer phase the leg loses its contact with the supporting surface and fails to participate in balance maintenance.

In the process of normal walking all the legs follow a cyclic pattern. The cycle's time T in all legs is believed to be the same. The leg's operation is characterized by one parameter: γ regime coefficient i.e. the relative duration of the leg's support. In other words, if for the total time cycle T the leg is in a support phase t_s, in this case

$$\gamma = t_s / T$$

In the process of walking different legs (not in the very same phase) are positioned dissimilarly in relation to the point of their suspension. For example, one leg may have merely lowered onto a supporting surface, whereas the other one may be completing its support phase and lifting. Hence, it is necessary to introduce a parameter featuring the position of the leg within the cycle, or, more precisely, the leg's relative position.

Owing to the fact that the cycle's time in all legs is alike the leg's position within the cycle may be characterized by the time passed from the moment of the beginning of the leg's cycle or the time remaining to its end. The moment of the support's phase shifting into the transfer phase is adopted as the moment of the cycle's beginning (*). The time left till the end of the cycle measured in twelfths of its cycle is characterized by the leg's position number, or for short - position. The position number may vary from 0 to 12 and the leg's position in 0 and 12 positions coincide; in an analogous way positions 13 and 1,14 and 2,....25 and 1 etc. also coincide.

It is evident that any of the leg's positions is recurrent in T time. Moreover, the difference of positions of any two legs in the process of walking is invariable. It is a time sequence of similarity in the operation cycle of each leg. A fixed state of any leg's position is denoted as p which stands for the initial position. The position of the right hind leg at the moment of its tearing away from the supporting surface i.e. the moment of its cycle (position 0) is regarded as the initial position.

The initial position of any leg at a prefixed position of the first (right hind) leg is a convenient parameter for the gait's description. In a general case any gait in six-legged vehicles is characterized by a set of 11 parameters: 6 regime

---

(*) As a rule, the moments of the cycle's beginning in different legs fail to coincide.

coefficients γ_i and 5 initial leg's positions p_j (i = 1, 2, 3.....6) as it is assumed that p_1 = 0 where i reflects the leg number read off in an counterclockwise direction beginning from the right hind leg.

Such a number of parameters is too great for analysis. The most natural way for reducing the number of parameters concurrently with simplifying the control system of a particular facility would be to establish a uniform regime for the leg's operation. Thus, it will be further adopted that $\gamma_i = \gamma$ (i = 1, 2......6). The limits of γ variations in a six-legged vehicle designed for statically stable walk are $1 > \gamma \geqslant 0.5$.

Had $\gamma = 1$ the leg would be transferred instantly, t_p = T and the time of transference would amount to 0, whereas when $\gamma < 0.5$ there turn out to be moments when only two legs are in touch with the earth and statical stability is impossible. It is worth noting that hexapod representatives of the animal world fail to exhibit various regimes in different legs [6]; symmetrical gaits are most rarely observed in quadruped specimens [5]. Thus, the system is composed of six parameters: γ, p_2, p_3, p_4, p_5, p_6.

As stated above the aim of our research was to find most stable gaits. We started off by dividing the entire group of gaits into two: stable and unstable ones. The unstable gaits, in relation to six-legged vehicles in the process of walking, are those in which the projection of the centre of gravity happens to be beyond the supporting polygon at any given moment of time.

The task of exploring stable gaits i.e. the determination of a required regime γ and the setting of initial positions of five legs was solved with the aid of a computer. In the course of solving the problem the reserves of stability were determined. The term "reserves of stability" covers the distance from the projection of the vehicle's centre of gravity onto the supporting surface up to the boundary of the supporting polygon measured along the vehicle's longitudinal axis. The reserve of stability of any gait within a cycle's range is variable. The minimal reserve of stability z within a walking cycle has been adopted as a quantitative measure of stable walking. It has been assumed that the best gait is that in which z attains its maximum. Six parameters of gaits exclude the possibility for a graphic representation of the range of stable gaits within a parameter's spacing.

The exploration of the best gaits within the range of stable ones has shown, firstly, a rise in stability with an increase of the coefficient's regime γ. This most natural conclusion follows from the fact that it is involved with an increase of the relative time of the leg's stay on the supporting surface. Secondly, optimal gaits reveal a number of features which permit the formation of a specific group of symmetric gaits. The notion "symmetry" is used in the sense adopted in biological literature and denotes a respective position of legs in the vehicle's right and left half shifted by a half a cycle. In this sense symmetry reduces the number of parameters to three: the regime coefficient γ and the initial positions of two legs, for example, the right mean p_2 and the right foreleg p_3. The initial position of the remaining legs is determined in the following way:

$$p_1 = 0, \ p_5 = \begin{cases} p_2 - 6, & \text{if } p_2 \geqslant 6 \\ p_2 + 6, & \text{if } p_2 < 6 \end{cases}, \ p_4 = \begin{cases} p_3 - 6, & \text{if } p_3 \geqslant 6 \\ p_3 + 6, & \text{if } p_3 < 6 \end{cases}, \ p_6 = 6$$

Three parameters permit us to draw up a visual graphic representation of the range of stability of symmetric gaits within the parameters' spacing. This is shown in fig. 1b, whereas fig. 2 displays sections of the stability range through $\gamma = 1/12$ intervals.

Attention ought to be paid to some features depicting stability ranges involved with the cyclicity of the chosen independent parameters p_2 and p_3. The choice of the origin of coordinates at point 0,0,0 as well as the very attachment of the cycle's initial point to the moment of the leg's lifting is arbitrary. However, due to the choice of the origin of coordinates the shape of the stability range represented on the plane changes its appearance.

This may happen providing the limits of parameter variation p_2 and p_3 are restricted by position numbers ranging from 0 to 12. For the sake of making things more pictorial it is convenient to extend the range of parameter variation p_2 and p_3 onto the second and even the third cycle as shown in fig. 2. Strictly speaking a representation projected onto a plane requires a display not of a single range, but of a series of merged ranges. A similar system of ranges is no longer dependent upon the choice of an initial point and may be regarded, to a certain extent, as an invariant one. It is shown in fig. 3. However, in this case the visual representation is inferior to the one indicated in Fig. 2 and 1b.

As mentioned above, the minimal reserve of static stability Z was adopted as a quantitative measure of stability. In the course of computing operations the rate of the stability reserve was calculated for all points of the stable range covering p_2 and p_3 variations from 0 to 12 and γ from 0.5 to 1 with intervals of 1/12. The analysis of these rates has shown that for each γ the maximal values of stability reserves are groupped about the main "diagonal" of the range - lines - ab (fig. 2).

The gaits, with their initial positions spaced along the main "diagonal", reveal a common property which may be depicted as

$$p_2 - p_1 = p_3 - p_2$$

In other words, the hind, mean and forelegs begin to raise at similar time intervals.

These gaits may be characterized as "wavy" ones, for the legs' operation is much like a wave distribution along the frame. This situation coincides with Wilson's idea (6) concerned with the spreading of an excitation wave in hexapod insects (and octapod arthropoda). All the "Wilsonian" gaits are spaced within the main ab diagonal of the stability zone.

There are two independent parameters in "wave" gaits (for example, γ and p_2, because $p_1 = 0$, $p_3 = 2p_2$ if $p_2 \leqslant 6$ and $p_3 = 2p_2 - 12$ if $p_2 > 6$.

Hence, the range of their stability may be reproduced on a plane (fig. 4). The range appears to be a projection onto the section's co-ordinate plane $(1 - \gamma)0p_2$) of the stability range built up by a vertical plane (fig. 1b) running through the main "diagonal". The very same fig. 4 displays "isostats" - lines connecting points of similar stability reserve. The range close to point b reveals maximal stability. The features of the isostat pattern exhibit an anomalous abc range of a lesser stability reserve.

And, indeed, had one followed up the mode of stability variation applied to a particular gait featured by γ decrease (fig. 5), one would observe an abrupt jump of stability reduction. This phenomenon may be ascribed to the fact that in range 11 fig. 5 stability is provided by the marginal and mean legs, whereas in range 1 there are moments when the support is achieved by the hind and forelegs. Owing to this the boundary of the supporting polygon runs much closer to the centre of the vehicle's gravity.

As to the order of the legs' operation for different gaits it may be stated that the particular case of symmetric gaits is represented in fig. 6; in the latter the lines of simultaneous legs' operation are traced onto the parameters' planes (p_2, p_3). The same figure also depicts the range of stability hatched for $\gamma = 10/12$. It is understood that the vehicle featured by $\gamma = 10/12$ is capable of performing the same set of gaits which fall within the hatched range.

With the change of γ their number will vary. For example, in the case of $\gamma = 0.5$ the number of stable gaits is but 3. One gait is achieved when simultaneously only one leg is raised at a time - "reserve transectional wave"; the other one is distinguished by moments when a pair of legs is raised at once, and, finally, the third is characterized by a simultaneous operation of three legs together - the so-called gait in "triples".

With an increase of γ the number of different symmetric gaits raises and at $\gamma = 1$ the number approaches the theoretical limit of 24. These gaits are represented in fig. 7 in which numerals indicate the leg number while letters correspond to the line points and the ranges on the parameter plane (fig. 6). The five lower gaits in fig. 7 (t, n, x, v, i) are possible only at $\gamma = 1$.

The order of the legs' operation is marked by arrows in fig. 8 and represent 8 chief types of gaits with single depicted leg operation.

In conclusion it is worth emphasizing once again that the stability of all gaits performed by six-legged vehicles with a similar regime in all legs was calculated by a computer for a definite interval. It has been shown that symmetric gaits, particularly wavy ones, exhibit maximal stability. It appears that in the course of designing six-legged walking apparatuses symmetric gaits will be regarded as guiding ones. For the purpose of achieving utmost stability at a determined regime γ_0 it is necessary to choose initial legs' positions close to the crossing point of the straightline $\gamma = \gamma_0$ and ab line shown in fig. 4.

REFERENCES

[1] GRAY J., Studies in the mechanics of tetrapod skeleton, Journ. Exper. Biol., 7, 1944.

[2] FRANK A.A., McGHEE R.B., Some considerations relating to the design of autopilot for legged vehicles, Terramechan., 6, 1, 1969.

[3] PETTERNELLA M., I veicoli camminatori, L'Elettrotecnica, LIX, 5, 1972.

[4] OHOTSIMUSKI D.E., PLATONOV A.K., Digital computer-simulation of the legged locomotion system, IV IFAC Symposium on automatic control in space, Dubrovnik, 1971.

[5] SUKHANOV V.B., General system of symmetrical locomotion of terrestrial vertebrates and some features of movement of lower tetrapods. "Nauka", Leningrad, 1968 (in Russian).

[6] WILSON D.M., Insect walking, Annual Rev. Entomol., II, 1966.

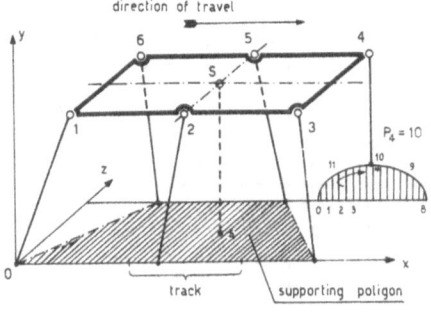

fig. 1a

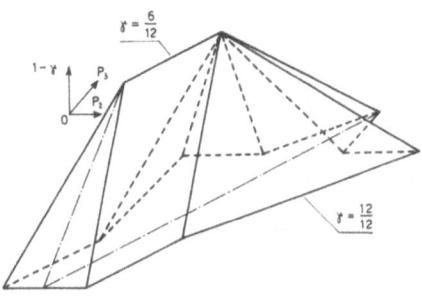

fig. 1b

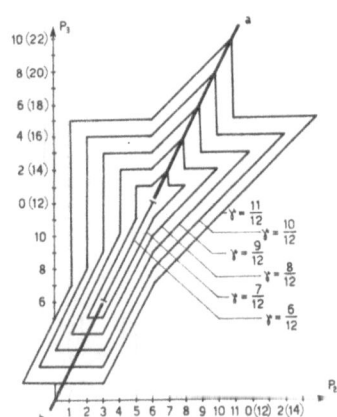

fig. 2

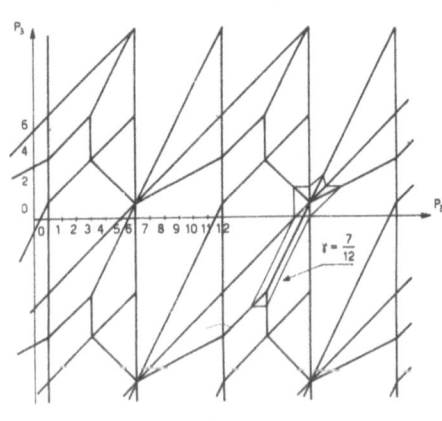

fig. 3

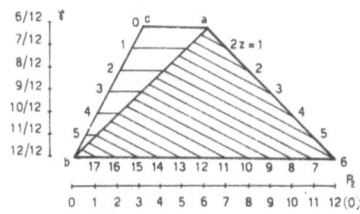

fig. 4

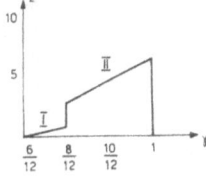

fig. 5

fig. 7

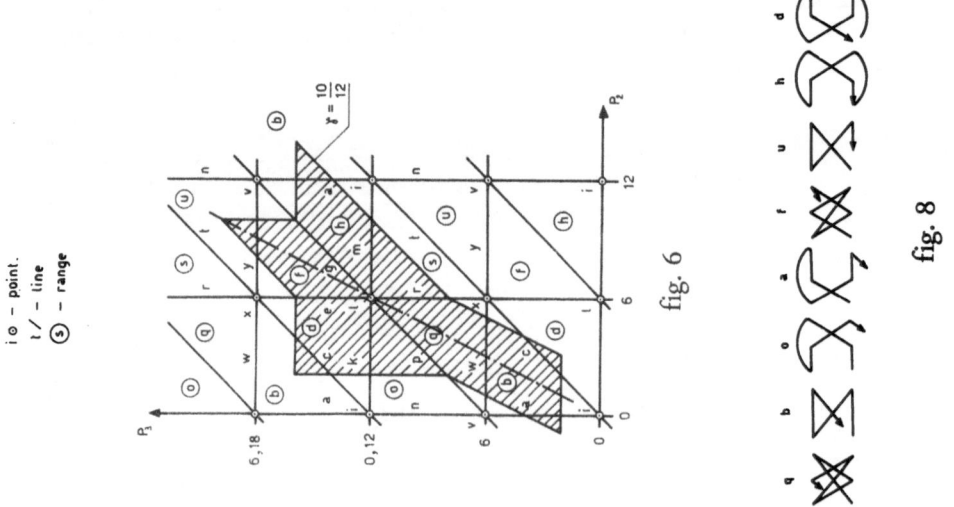

fig. 6

fig. 8

INFORMATION—POWER MACHINE WITH SENSES AND LIMBS
(Wabot 1)

Ichiro KATO, Professor,
University of Waseda, Department of
Mechanical Engineering, Tokyo, Japan

Sadamu OHTERU, Professor,
University of Waseda, Department of
Applied Physics, Tokyo, Japan

Hiroshi KOBAYASHI, Professor,
University of Waseda, Department of
Applied Physics, Tokyo, Japan

Katsuhiko SHIRAI, Associate Professor,
University of Waseda, Department of
Electrical Engineering, Tokyo, Japan

Akihiko UCHIYAMA, Professor,
University of Waseda, Department of
Electronics and Communications,
Tokyo, Japan

(*)

(*) All pictures quoted in the text are at the end of the lecture

Summary

    The WABOT project is aimed at developing an information-power machine to introduce external and internal informations, judge the same by itself and perform a given action autonomously. The WABOT-1 consists of a human-type biped walking machine provided with artificial ears and mouth to receive a vocal command from the operator and make a vocal response, artificial eyes to recognize the position, distance and direction of an object and human-type bilateral hands. Under the control of a mini-computer as the artificial brain and on the basis of external informations received from the visual, auditory and dermal senses, etc., the WABOT-1 performs a given work in linked motions of hands and feet in on-line real time.

Zusammenfassung

    Theoretische Untersuchung zweibeiniges Gehens is notwendig für Erkentnisse von menschlichen Gehen und für Konstruktion von künstlichen Beinen und von Gangmaschinen. Dynamische Stabilität ist eine Hauptfrage in diesem Gebiet.

    Eine Methode zur Stabilitätsregelung einer Gangmaschine wird vom dynamischen Gesichtspunkt aus behandelt. Unter der Voraussetzung, dass die Maschine mit einem festen Körper und zwei Beinen ohne Masse ausgerüstet ist, wird ihre Bewegung im Hinblick auf die Reibungskraft zwischen dem Fuss und dem Boden untersucht. Die Regelungsmethode kann ein der beim menschlichen Gehen auf der Ebene erwiesenen Charakteristik gleichendes Bewegungsbild darstellen.

1. Advent of Robot

The Group of Bio-Engineering in Waseda University has been carrying out a robot project these five years. Principal particulars of the specification are as follows:

A. To move on a human-type biped walking machine.

B. To work with bilateral artificial hands of a human type.

C. To provide hands and feet with joint angle sensors for each joint as proprioceptive means.

D. To provide a sense of equilibrium for a biped move.

E. To provide hands with contact sense for work.

F. To provide artificial eyes as remote receptors.

G. To use a minicomputer as the brain.

H. To speak and listen Japanese sentences.

We named the robot "WABOT (WASEDA ROBOT)-1". Fig. 1 shows the complete figure of Wabot-1.

2. Construction of Biped Walking Machine WL-5

The basic construction of the machine model is a joint construction of a revolving movement type, with each being of a hinge structure with one degree of freedom (Fig. 2). The machine model possesses 11 joints i.e. two plantal joints to incline the whole feet to the left or right for shifting the center of gravity to the left or right, two ankle joints, two knee joints, two hip joints, two loin joints for direction change and one body joint to incline the upper part of body to the left or right to help a lateral shift of the center of gravity. Hydraulic pressure is employed as a drive source and hydraulic circuit components except a power unit are built in. Further, the actuator is a cylinder of a direct-acting type. The weight of the machine proper is approx. 130 kg, and a meter-out system (and partially a meter-in system) is employed in the hydraulic circuit with a working pressure of $70kg/cm^2$.

Each joint employs a rotary type of potentiometer to form an angle control mechanism. This mechanism is controlled in applying the gait pattern, which corresponds to a continuous walking behaviour and is divided into some characteristic phases. By realizing such discontinuous phases successively by the angle control mechanism, one gait pattern is performed. This gait pattern is stored in CPU as digital data and gives out each phase one after another when an order for walking is given from outside (man or other subsystems: eyes and ears). The components of the system and their operations are explained as follows. Upon an order for walking, a phase stored in the first address is given out from CPU. (Fig. 3).

CPU digital output unit: Within CPU, one phase is stored in one address and a collection of such phases constitutes one gait pattern.

Discrimination circuit: Each phase is a set of angle values to be taken by each joint. The value of the aim angle is D/A-converted and compared with the value (obtainable from a potentiometer) of the existing angle of each joint.

Mechanical model: Depending on whether the error is positive or negative, the cylinder expands or shrinks and as a result, each joint reaches an aim angle required by the phase.

Detecting circuit: The phase is realized when all the joints have reached those aim angles. A pulse is given out at that time.

CPU digital input: If the said pulse enters CPU, the address advances by one step and the next phase stored in the second address is given out.

By achieving a series of phases successively in this way, one gait pattern is accomplished and walking is realized. Further, in any phase during such walk, the center of gravity of feet falls within the sole in a standing phase and therefore a statically stable walk takes place at all times.

3. Human-type Bilateral Artificial Hands

The manipulators are of a human type and same in construction for left and right hands. Each hand has 7 degrees of freedom as in Fig. 4, comprising shoulder rotation, upper arm rotation, elbow revolution, wrist rotation, vertical wrist movement, horizontal wrist movement and opening and closing of fingers. The hand region comprises 5 fingers of a linked joint structure. Each joint is provided with a DC motor by way of a reduction gear.

As the external information detectors, microswitches acting as tactual sensation, 8 per hand, are provided in the hand region. Six of them for fingertips and palm are used for detection of a gripping condition and connected to the computer by way of a logic circuit. Signals from search microswitches, two per hand, fitted to the left and right ends of the hand region become interruption signals after passing the logic circuit.

As for the internal information detectors, potentiometers are provided for angle detection of the respective drive joints. However, the condition of opening and closing of fingers is detected by a microswitch.

Concerning the software, a program system of a hierarchical structure has been developed to simplify the orders to be given by the operator. The program comprises:

(1) A control program

(2) A basic work program module group

(3) A basic program module group

(4) A typewriter-control main program

(5) A typewriter-control subprogram module group

The control program is a program intended for causing artificial hands to perform work by controlling a basic work program.

The basic work program module group is designed for causing action as the basis of work to be done and comprises a movement routine, a search routine and a finger closing-opening routine, etc.. The movement routine is so designed that in case where the present point of hand and the target point for movement are given in three-dimensional rectangular coordinates and the hand posture in a direction cosine, it calculates transit points dividing a straight route from the present point to the target point in 8 equal parts in terms of the three-dimensional rectangular coordinates by means of a transit point calculation routine, converts the three-dimensional rectangular coordinates of the next transit point into each joint angle by means of a coordinate conversion routine and thereafter makes a move to the next transit point by means of a basic movement routine and in repeating the above processes, achieves a move to the target point via a straight route. In this case, if a calculated transit point is not included in the working range of the artificial hand, a correction is made by a transit point correction routine so as to pass a point within the working range.

In the search routine, first, if a search commencement point is given in the form of three-dimensional rectangular coordinates, one of the left and right hands is selected by left-and-right hands selection routine and a search posture is data-transferred. Each joint angle at the search commencement point is calculated by the coordinate conversion routine from the search commencement point given in the form of three-dimensional rectangular coordinates. If, in this case, the search commencement point falls outside the working range of the artificial hand, stop takes place after giving out a message to that effect. If the search commencement point falls within the working range of the artificial hand, move to the commencement point is made by the basic movement routine and search is commenced by a horizontal movement routine. If, during a horizontal movement, the artificial hand touches the object, a search microswitch provided in the hand region is turned on, applying interruption to CPU. After analyzing it by means of an interruption analyzing routine, touch of the artificial hand with the object is confirmed by a sensor utilization routine, to stop the artificial hand.

The basic program module is a minimum routine constituting the basic work program and comprises a coordinate conversion routine, a basic movement routine , a horizontal movement routine and a transit point calculating routine, etc..

The typewriter-control main program is designed for control of the artificial hand though conversations between man and the computer by way of a typewriter. The typewriter-control subprogram module causes basic actions required for control of the artificial hand through the typewriter to be performed and is controlled by the typewriter-control main program. The typewriter-control subprogram module comprises a memory executing routine and a left-and-right movement routine, etc..

4. A system for Processing Visual Data

Wabot has its eyes constructed by the two TV Cameras in its trunk, whose detector scanning lines and focusing are set by the CPU control. When searching for an object, Wabot's trunk rotates to scan the space in front with its TV Cameras.

When the scanning camera detects an object, it emits a signal which stops the trunk rotating. All data on distance and angle is detected and processed at the same time, as an 8-bit digital value. If the eyes don't detect an object, scanning lines are shifted and the detecting procedure is repeated following the software algorithm. In order to improve the accuracy of measurements, the zone focusing with a 3-step on-off control and the broadening of the visual angle for a short distance are added in the mechanism of TV Camera.

The measuring error of distance is within 7 cm at 5m, and 4cm at the shortest possible working distance of 0.6m.

Calculations of distance and angle and interrupted processing from the hardware are controlled by the software. Wabot's Eyes are built in the body of Wabot as seen in Fig. 1. Fig. 5 shows the interface of the trunk and the eyes.

In order to first catch the object within visual range, the video signals from the object are picked up by both TV Cameras (Camera I, Camera II), using the line selector for the vertical direction and the trunk rotation for the horizontal direction. The gray level of the object signal is sliced and transformed into a binary value. The jumping position of this binary signal in the Camera is counted with a 4MHz oscillator, and the gate is closed through the hold logic. When, with the rotation of the trunk, the signal made by each Camera is detected at the equal distance from the edge of its sight, the rotation is stopped. At this time, since the trunk is facing the object directly, the data for the rotating angle between the legs and the trunk to be sent to the legs is obtained, and thus the walking direction is determined. Since the distance between two cameras is known beforehand, the data for the distance to the object is also determined.

The following is a summary of the essential points to be considered with regard to the specification of the eyes of this 2-legged robot;

(1) Since the robot walks by shifting its center of gravity, many problems arise in detecting an object while walking.

Therefore Wabot's eyes were designed this time to function only when not moving.

(2) Even when Wabot is still, the detector scanning lines of the TV Cameras cannot be fixed even to a stationary object, because of the sight movements made in both vertical and horizontal axes. In the trial system, 64 different scanning lines were made available for selection, and one of them was selected by the software.

(3) The walking direction as well as the distance must be determined prior to starting to walk, since it is necessary to set the robot in the right direction facing the object.

(4) Since a mini-computer is used, data is processed by hardware as much as possible.

5. Speech-Input-Output System

The Speech-Input-Output System recognizes voice commands to the WABOT and makes vocal responses to the operator. It involves the problems of pattern recognition, understanding of meanings and speech synthesis by computer (Fig. 6).

The circumstances around the WABOT and the jobs which it can perform are limited in this case. The command sentence to the WABOT consists of a string of Japanese words which are separately spoken with pauses between them. Japanese sentences have a little flexibility in the order of words which, compose them, so that the system is designed to recognize the differently ordered sentence for the same meaning.

If some words cannot be recognized or the command is not appropriate for the situation the system makes the vocal response such as not understandable , inadequate or impossible and it asks to operator to repeat the command sentence. If an imperfect sentence in meaning was spoken, it asks a question to make the sentence completed. And when the voice command is accepted, it repeats the command by the speech synthetizer and then produces the command codes for the control system to move the WABOT.

(1) Recognition of Speech

Speech sounds (voice commands) are at first preprocessed after catched by a microphone and the features are extracted by the hardware feature extractor. They are fed into a minicomputer in which they are used to generate a string of phonemes by the phoneme recognition program. And the string of phonemes is put into the word recognition program where the values of discriminant functions for the words of candidates are calculated. The discriminant functions are linear functions with variables which represent the transition of phonemes in a string, and the coefficients of which are estimated through a learning process. Because of learning ability, the vocabulary which can be recognized are easily changed.

(2) Understanding of Meaning

It is difficult to treat generally this problem. Therefore in this case the circumstances have been restricted to the situation that the WABOT walks the specified steps, turns round to right or left and stops according to commands.

The command sentences are composed of a string of words, STOP, BEGIN, WAIT, WALK (GO), TURN RIGHT, LEFT, ONE STEP, TWO STEPS, and THREE STEPS. These words are grouped into four classes according to their parts of speech and their meanings, which are VB1, VB2 (verbs), AD1, AD2 (adverbs). Possible sentences from these words are also grouped into six classes according to their structures.

To make it easier to properly recognize what sentence is spoken and what is the meaning, three types of states of the WABOT are introduced. They are:

ST1 : the state waiting for a command,

ST2 : the state after it recognized a command and waiting for the final instruction, "wait" or "begin", and

ST3 : the state carrying out a job and waiting for the command to stop or no command.

In deciding the kind of the sentence was spoken and the meaning, a dynamic programming technique is used.

(3) Synthesis of Speech

The terminal analog synthetizer is used to make vocal responses. The method of synthesis by rule is adopted. A word is made of a string of phonemes, each of which has the specified pitch, length and strength.

6. Central Processing Unit and Interface

Central processing unit (CPU) is a minicomputer with 8k words (16 bits/word) core memories, while interface has been designed to relay all signals between CPU and various subsystems of the WABOT.

(1) Hardware

The interface blockdiagram is illustrated in Fig. 7.

Analog input (AI) includes multiplexer (MPX) sampling and holding circuits, A–D converter (12 bits) and decoder to analyze a control signal from the CPU.

Digital input (DI) part is constructed from data register, interruption register, channel register, channel decoder, timing pulse generator and order decoder to analyze various instructions from the CPU.

Analog output (AO) contains two D–A converters, output gates, instruction decoder. These D–A converters have 12 bits buffer register for each signal.

Digital output (DO) is composed of data register, channel register, channel decoder, timing pulse generator and order decoder to analyze IOC instructions.

Timer generates timing pulses to start A–D conversion, D–A conversion and to control WABOT systems in constant time interval.

(2) Software

Special operating system program (WABOS) is prepared for WABOT use.

WABOS is divided into the following six programs.

(a) Loader

Loader includes two types, WABOL (1) and WABOL (2).

The former loads a task registered by scheduler to core memory, while the other loads a task by interruption which is not in core memory, during program running state.

(b) Scheduler

Scheduler (SCH) includes two types, SCH (1) and SCH (2).

SCH (1) regists various tasks prior to program performance and SCH (2) passes and receives program control during program running state.

(c) Interrupt analysis routine

The routine analyzes the origin of interruptions.

(d) Interrupt service routine

The routine analyzes every interruptions from input output devices and those from CPU in order to service these request, that is, if the interruption is the one from 10 devices, the routine looks a table (Event control table) and rewrite the table, then transfers control to scheduler (1).

(e) Macro-service routine

Two types of macro-instructions are prepared. There are eight 10 macro-instructions and seven data request macro-instructions.

10 macro-instructions are, for instance, transfer instruction that moves the contents of the accumulator to output data register, read instruction of analog data to the accumulator, etc..

Data request macro-instructions, for instance, CALL instruction is utilized to call subroutines under OS supervisory, END instruction that ends all operations on main task.

(f) Function subroutine

The subroutines are divided into two groupes, basic function subroutines and conversion subroutines.

7. Conclusion

At the present stage because of the restriction of CPU memories the total operation is not yet made. However, the following experiments have been completed for each subsystem:

(A) Foot subsystem

After loading with a load of 20 kg, the machine model was able to perform the various walking a straight walk and a direction change to the left or right under the control of CPU.

(B) Hand subsystem

In an example of work, the machine model receives an information on the position of an object on a desk from the visual information processing system via CPU, searches the vicinity and completes a searching work when the hand microswitch acting for a tactual sensation touches the object and is turned on, grips the object and passes it into the other hand, after which it moves to the garget point.

(C) Eye subsystem

An experiment was conducted by manually setting a command from the computer to the eye interface mechanism, in which the object detection mechanism, angle detection mechanism and WABOT torso control mechanism (TV Camera focus control , sight extension control, torso rotation control) were able to carry out required actions.

(D) Voice subsystem

By combining the speech input-output system and the foot subsystems it was possible to perform limited actions by voice commands.

ACKNOWLEDGEMENTS

The authors wish to thank the Group of Bio-engineering of Waseda University for their cooperation.

The work was financially supported in part by the Group project organized by Science and Engineering Research Laboratory Waseda University and Grant-in-Aid for Fundamental Science Research from Japanese Ministry of Education.

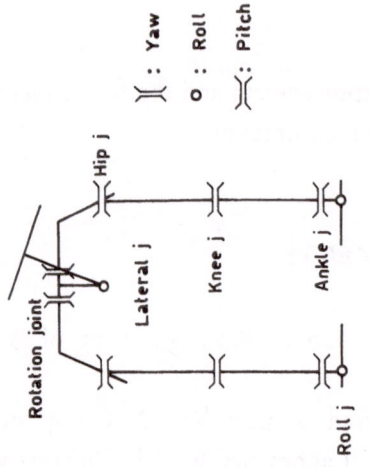

fig. 2 Degrees of freedom of WL-5

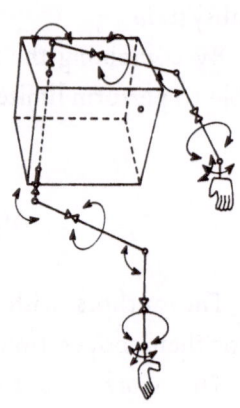

fig. 4 Degrees of freedom of WAM-4

fig. 1 Wabot-1

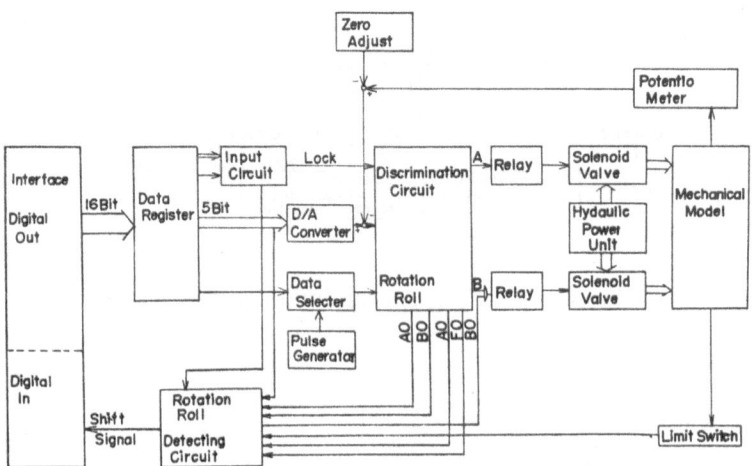

fig. 3 The control system of WL-5

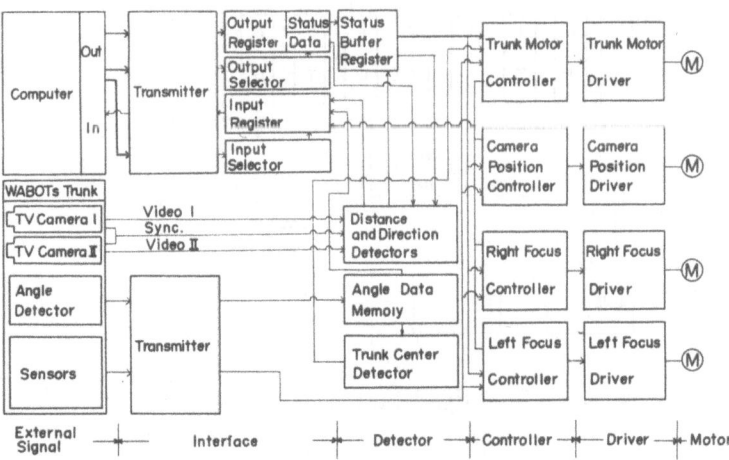

fig. 5 Block diagram of Wabot's eyes and their control system

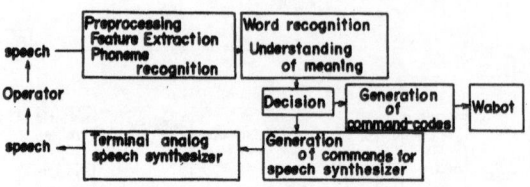

fig. 6 Block diagram for speech-input-output system

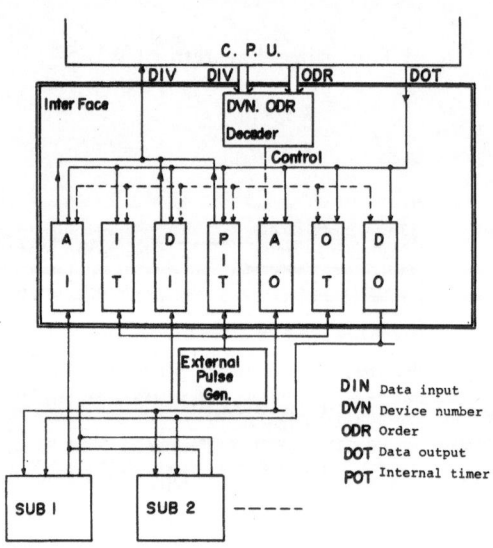

fig. 7 Interface

AN INTERACTIVE COMPUTER–CONTROL SYSTEM FOR A QUADRUPED ROBOT(*)

R.B. McGHEE and D.E. ORIN
Department of Electrical Engineering, Ohio
State University, Columbus, U.S.A.

(**)

Summary

While the mobility advantages of legged locomotion systems relative to wheeled or tacked systems are generally recognized, the complexity of the joint coordination control problem has retarded the development of useful legged vehicles. This paper reports the results of a computer simulation study in which a digital computer was used to generate commands to the individual joints of a quadruped robot in response to speed and direction signals from a human operator. The paper includes a detailed exposition of the necessary calculations and a discussion of operator reaction to the proposed control scheme.

(*) This research was supported by the National Science Foundation under Grant GK-25292

(**) All figures quoted in the text are at the end of the lecture

Introduction

It is generally recognized that terrestrial animals, including man, enjoy superior mobility in comparison to wheeled or tracked vehicles when moving over difficult terrain or in restricted spaces. Typically, artificial systems based on the wheel offer advantages only for locomotion over relatively hard and smooth surfaces [1, 2]. This observation has motivated a number of investigations into the possibility of realizing vehicles making use of systems of levers rather than wheels for support and propulsion. Successful automatic coordination of joint motions in such "legged" locomotion systems has been reported for machines with two, four, six and eight legs [2, 3, 4, 5, 6, 7]. In addition, the "master-slave" approach to manual control of remote manipulators has been successfully adapted to artificial bipeds [3] and quadrupeds [8].

Generally speaking, up to the present time, automatic systems for vehicle limb motion coordination have been characterized by fixed gait patterns with little or no ability to adapt to terrain variation or to changes in route, while manually controlled systems have burdened the operator with direct control of every joint of each limb. Neither of these approaches has succeeded in producing a vehicle which is at all comparable to natural systems with regard to speed, mobility , or efficiency of locomotion. It now appears that it may be desirable to combine these two approaches through the technique of "supervisory remote control" [9] in which a type of man-machine symbiosis is sought by assigning to a human operator the task of making complex decisions such as those relating to selection of destination, specification of route, speed, etc., while relegating to a computer the relatively simple but burdensome task of generating commands to individual joints. This approach has been found to be very effective in manipulator control [10, 11] . The present paper is devoted to an exploration of the potential of interactive computer-control in the context of a simulation study of a quadruped robot.

Quadruped system description

Among automatically controlled legged locomotion systems, it is possible to distinguish between linkage machines [2, 6, 12] with only one or two mechanical degrees of freedom, and electronically coordinated machines [3, 4, 5, 7] in which every joint of every limb is capable of independent movement. Due to the limited adaptability to a changing environment inherent in machines with reduced degrees of freedom, only the latter class of systems will be treated in what follows. Furthermore, since a large man-carrying manually-controlled artificial quadruped exists [8] and because one of the authors has previously investigated certain aspects of both natural and synthetic quadruped locomotion [4, 13, 14, 15, 16], the supervisory remote control system to be described in this paper is specialized to quadrupeds. It is believed, however, that the ideas presented are general enough to permit application to other types of legged locomotion systems.

Because computer simulation studies have previously been found to be very useful in the design and analysis of legged locomotion systems, especially in early stages [16, 17, 18], it was decided to study quadruped control by means of a simulated rather than a real vehicle. Experience with this approach to locomotion studies [16, 18], and with experimental investigations [4, 19], has given the authors confidence that simulation results can be translated into hardware performance providing that the idealizations inherent in simulation are adequately taken into account. Accordingly, the dynamic simulation described in [16] was used in place of a real vehicle to test the computer control system. Moreover, the "model reference" control scheme described in [16] was also adopted. This approach leads to a partitioning of the control problem into a kinematic part and a dynamic part as illustrated in Figure 1.

The kinematic part of model reference control involves interpretation of human operator inputs to determine desired angles and rates for each joint of the vehicle support system. For the system under discussion, this function is accomplished by a set of computer programs which have been collectively labelled the kinematic command generator [20]. As shown on Figure 1, the output of the kinematic command generator is a vector, v, which in turn serves as input to another computer program module called the force and moment generator. The latter program determines the joint torques and leg forces needed to cause the vehicle to closely approximate the behavior specified by v. The force and moment generator used in this simulation study is described in complete detail in [16] and [18]. The remainder of the present paper is devoted to a description of a particular kinematic command generator for quadruped control [20]. It is hoped that the details presented will be helpful to others interested in computer-control of legged locomotion systems.

Kinematic command generator

Referring to Figure 1, note that the inputs to the kinematic command generator are provided by an operator. With visual and possibly other physiological feedback as to the present state of the vehicle and the terrain, the operator provides commands related to the desired velocity and direction of the vehicle. For the system under discussion, there are three such inputs and they are:

1) Desired Velocity: The operator chooses an appropriate value for the desired velocity, VC, to be used to move the vehicle over the terrrain being covered. This velocity may be either positive or negative. For positive velocity, the direction of the desired motion is assumed to be that of the present longitudinal axis of the quadruped system while for negative velocity a rearward motion is implied.

2) Curvature: This quantity, KC, is used to determine the desired rate of turn of the vehicle with positive values corresponding to a turn to the right.

3) Stop-Start: The "stop" command generates a desired velocity of zero and proceeds to stop the vehicle. No other command as to speed and direction can be executed until a "start command is given.

Gait Selection

Four gaits are employed by the simulated quadruped. These are the crawl, the walk, the trot, and the gallop [4, 14, 15, 16, 21]. As in living quadrupeds, the crawl and the walk are used for low speed travel, the trot for intermediate speeds, and the gallop for high speed.

The selection of gait is not manually controlled by the operator in the present program; rather, there is an automatic shift from gait to gait at specified values for the velocity of the vehicle. Figure 2 shows the assumed relationship between velocity and gait. As can be seen, a 2ft/sec hysteresis in velocity has been utilized to eliminate any undesirable rapid changes in the gait selected. It may be noted that the following gait selection principle holds:

1) For several of the velocity intervals, the value of the velocity uniquely determines the gait selected.

2) Gait selection for the velocity intervals of 6.0ft/sec $\leq$ VEL $\leq$ 8.0 ft/sec and 30.0 ft/sec $\leq$ VEL $\leq$ 32.0 ft/sec is uniquely determined by the gait previously employed before the velocity entered the interval.

All of these choices for switching velocities are, of course, arbitrary. A careful study of the dynamics of a given vehicle would undoubtedly result in other values for gait switching velocities, but the principles involved in automatic gait selection would presumably be unchanged.

Gait Transition

Gait transitions occur when the gait selected by the automatic shift equations is other than the present gait. No gait transition takes place as the quadruped is halted. Rather, the quadruped is in the crawl gait as it stops and the only change from a normal crawl is that the quadruped halts with its feet directly under its hips.

A smooth transition between the crawl and walk occurs by changing the duty factor [14] for each leg continuously. The following relationship is used for this purpose:

$$\beta_i = .9167 - .2167 \frac{\lfloor VEL \rfloor}{6} \tag{1}$$

where $0 \leq |VEL| \leq 8$. Thus, in the crawl gait, the duty factor is largest while the quadruped is moving at very low speeds. As the speed increases, the duty factor decreases, and the quadruped goes smoothly into a walk.

The early photographic studies of Muybridge [21] relative to horses reveal how this animal makes the transition between the walk and trot and the transition between the trot and gallop. Figure 3 uses gait matrices [4] to show the various support patterns that occur in the transition between these gaits in horses. The transition between gaits involves an alteration of the phase relationships between the legs as well as the duty factor of each leg in a set of discontinuous steps. As indicated, this tansition is usually completed in about 2 or 3 gait periods.

Velocity

The interpreted (output) value of the desired velocity, VCI, is a function of the commanded velocity, VC, and the present gait employed by the quadruped, QSTATE. That is,

$$VC1 = f(VC, QSTATE) \tag{2}$$

VC is equal to zero if the operator has issued a stop command. Otherwise, VC is equal to the value of desired velocity as supplied by the operator.

If the gait to be selected as anticipated by the value of VC is the same as the present gait, then

$$VC1 = VC \tag{3}$$

Otherwise, a gait transition may be expected. During such a transition, the value of VC1 is held constant so as to allow the transition to occur under relatively stable conditions.

Path Cruvature

The interpreted value of the curvature, K, is exactly equal to the value,

KC, provided by the operator and may be positive or negative. The radius of curvature, R, is also a signed quantity and is related to the curvature by the relation

(4) $R = 1/K$

Kinematic Reference Model

The kinematic reference model is based on several assumptions. These are:

1) The center of gravity of the idealized reference model proceeds at a constant height above the ground. That is [16]

(5) $z_E = ZR = $ constant.

2) The ground is level.

3) Coordinated turns result by banking; i.e., the body moves to the inside of the turn and rolls about its translational axis such that its acceleration vector points to the center of the support pattern.

4) The stride length, STD, for all gaits is constant while the period, PER, varies according to the relation

(6) $VEL = STD/PER$

5) The feet are placed regularly at specified intervals in both time and space. The initial foot positions are calculated to lie on a circular track whose radius of curvature is approximately equal to the radius of curvature of the path followed by the center of gravity. The effective stride length of the legs on the inside of the turn is smaller than those on the outside so that the number of strides for each leg will be equal as the center of gravity moves through a specific distance along a circular arc.

The combined effect of the above assumptions on the quadruped vehicle kinematic behavior is illustrated in Figures 4 and 5.

Acceleration

The interpreted value of the velocity, VC1, serves as the steady-state value that the actual velocity of the vehicle, VEL, approaches. If there is a difference in the actual velocity and the desired velocity, then the vehicle accelerates or decelerates to remove the difference. This scheme, much like that employed by automobiles, is described by the following relationship linking VC1, VEL, and acceleration, $\dot{VEL}$:

$$\text{VEL} + k_1 \, (\text{VÉL}) = \text{VC1} \tag{7}$$

Banking

   Figure 4 shows a rear view of a quadruped vehicle turning in a coordinated manner. The indicated angle ϕ_c by which the vehicle rolls about its longitudinal axis to accomplish proper alignment of its total acceleration vector is evidently given by

$$\phi_c = \tan^{-1} \, ((\text{VEL})^2 \, K/g) \tag{8}$$

The commanded body roll rate, $\dot{\phi}_c$, can be obtained by differentiating this expression with the result

$$\dot{\phi}_c = \left(\frac{\cos^2 \phi_c}{g} \right) (2 \, (\text{VEL}) \, (\text{VÉL}) \, K + \dot{K} (\text{VEL})^2) . \tag{9}$$

From Figure 4, the lateral distance by which the center of gravity is offset from its normal position as a result of banking is evidently

$$\text{LD} = \text{ZR} \tan \phi_c . \tag{10}$$

Foot Position

   For a given path curvature, K, the sequence of foot prints for each leg is calculated to lie on a circular track. For each leg, the distance from the center of gravity to a radius vector intersecting initial foot position is determined from a constant arc length increment, s_i , given by

$$s_i = (a_i \pm (\beta_i /2) \, \text{STD}) \tag{11}$$

where a_i is the longitudinal coordinate of hip socket i in body coordinates [16] . The + sign in (11) is for positive velocities and the -sign is for negative velocities.

   Figure 5 shows a top view of a quadruped moving in a circle of constant radius. The anticipated sequence of foot positions for the front legs are shown. The radii of the circles containing foot positions are

$$R_i = [(R + b_i)^2 + a_i^2]^{1/2} \tag{12}$$

where (a_i , b_i) is the location of hip socket i in body coordinates [16] , i = 1, 2, 3, 4. For any leg, the next foot position can be determined geometrically by marking off a constant arc length s_i on the circular path of the center of gravity. A line bc is then drawn from the center of curvature c to point b. The point of intersection d between line bc and the circle whose radius is R_i marks the point of the next initial foot position for leg i. The x-y (body) coordinates of the next initial foot position for leg i are this determined by

(13)
$$
\begin{bmatrix} x_i \\ y_i \end{bmatrix} = P_1 \begin{bmatrix} R \\ \rho - R_i \end{bmatrix}
$$

where

(14)
$$
P_1 = \begin{bmatrix} \sin(s_i/R) & -\sin(s_i/\rho) \\ 1 - \cos(s_i/\rho) & (R/\rho)(\cos s_i/\rho) \end{bmatrix}
$$

and

ρ = absolute value of radius of curvature, R.

(x_i, y_i) = initial position of foot i expressed in body coordinates.

These expressions follow by noting that the x-y components of the distance from the center of gravity of the body to point d are just the sum of the longitudinal (x) and lateral (y) components from the center of gravity to point b and from point b to point d.

The above equations determine foot position and body position for locomotion over level ground in response to operator commands. The rest of the kinematic command generator program is concerned with conversion of this information into joint angle commands. The necessary equations for this purpose can be found in [16], [18] and [20].

Computer simulation results

In the preceding analysis, a computational procedure for a kinematic command generator has been developed. When incorporated into a computer program [20], this generator receives commands from a human operator and outputs joint angle commands. The kinematic command generator has been simulated on a Digital Equipment Corporation PDP-10 computer system. A two-degree-of-freedom joy stick was chosen to furnish velocity and directional commands. Forward (backward) motion of this joy stick corresponds to positive (negative) values on the velocity of the quadruped vehicle. Lateral motion of the stick to the left or right permits operator control of path curvature. In addition to the velocity and directional commands furnished by the joy stick, commands as to starting and stopping are entered through the teletype keyboard.

In order to permit a preliminary subjective evaluation of the potential of supervisory remote control in vehicle joint coordination, the human operator involved in this simulation study was furnished with a visual image of an idealized quadruped robot on a computer-controlled cathode ray tube display. Figure 6 illustrates this arrangement and also provides some examples of typical cathode ray tube images. By observation of the motion of the simulated robot over a checkerboard pattern on the display, the operator is furnished with an immediate visual indication of the effect of his control actions. This type of feedback was found to be very realistic and operator reaction was quite positive. After the display and control system programs were correctly operating, several randomly chosen individuals were asked to take control of the simulated system after a short demonstration. In every case, learning time was almost zero. The operators all felt that the sensation was essentially that of remotely controlling an automobile or other wheeled vehicle.

While the basic purpose of the present investigation was to explore the potential of supervisory remote control in the generation of joint angle commands, and the results shown on Figure 6 relate only to kinematics, some experiments were also made in which the force and moment generator was included along with simulated vehicle feedback signals to produce a full implementation of the system of Figure 1. It was found that the proposed partitioning of the control problem into its kinematic and dynamic aspects was very effective in the crawl and walk gaits. For these gaits, the system was found to be quite insensitive to large variations in the feedback gains of the simulated joint control servomechanisms. However, when higher speed gaits were attempted, instability resulted and the simulated system "crashed". Preliminary indications are that stabilization of high speed locomotion may require some type of "step control" [22] in which the vehicle footprint sequence is not determined purely by operator input, but also involves some feedback of the vehicle dynamic state to the kinematic command generator.

Conclusions

This paper describes in some detail a specific computer implementation of supervisory remote control for a quadruped robot locomotion system. Many arbitrary choices of control techniques and parameters were involved in this study and none of these choices is believed to be optimal. However, the results obtained are encouraging and seem to warrant further investigation. Joystick control of vehicle motion is certainly very much easier than manual control of individual joints. The computer programs needed are not complicated [20] and could easily be implemented either on a small minicomputer or as a special purpose computer.

While much has been left undone, the authors hope that this paper will stimulate others to conduct further studies relative to the utility of computer control of joint motions, both in the context of robot locomotion systems and relative to assistive systems for the disabled. Research involving actual hardware studies would be particularly valuable at this point in time.

REFERENCES

[1]   Bekker, M.G., Off-The-Road Locomotion, University of Michigan Press, Ann Arbor, Michigan, U.S.A., 1960.

[2]   McKenney, J.D., "Investigation for a Walking Device for High Efficiency Lunar Locomotion," Paper 2016-61, American Rocket Society, Space Flight Report to the Nation, New York, October 9-15, 1961.

[3]   Kato, I., and Tsuiki, H., "Hydraulically Powered Biped Walking Machine with a High Carrying Capacity," Proceedings of Fourth International Symposium on External Control of Human Extremities, Aug. 28 Sept. 2, 1972, Dubrovnik, Yugoslavia.

[4]   McGhee, R.B., "Finite State Control of Quadruped Locomotion", Proceedings of Second International Symposium on External Control of Human Extremities, Aug. 29-Sept. 2, 1966, Dubrovnik, Yugoslavia.

[5]   Mocci, U., Petternella, M., and Salinari, S.,"Experiments with Six Legged Walking Machines with Fixed Gait", Report 2-12, Institute Of Automation, University of Rome, June, 1972.

[6]   Baldwin, W.C., and Miller, J.V.,"Multi-Legged Walker, Fin Report", Space General Corporation, El Monte, California, U.S.A., Jan. 30, 1966.

[7]   Vukobratovic, M., Ciric, V., Hristic, D., and Stepanenko, J., "Contribution to the Study of Anthropomorphic Robots," Paper 18.1, Proceedings of IFAC V World Congress, Paris, June, 1972.

[8]   Mosher, R.S., "Exploring the Potential of a Quadruped", Paper 690191, Society of Automotive Engineers, International Automotive Engineering Congress, Detroit, Michigan, U.S.A., January 13-17, 1969.

[9]   Ferrell, W.R., and Sheridan, T.B., "Supervisory Remote Control of Remote Manipulation", IEEE Spectrum, pp. 81-88, October 1967.

[10]   Whitney, D.E., "Coordinated Motion Control of Prosthetic Arms and Remote Manipulators", Paper 70-Mech-75, American Society of Mechanical Engineers, Mechanism Conference, Columbus, Ohio, U.S.A., November 1-4, 1970.

[11]   Maric, M., Gavrilovic, M., Radovanovic, D., "Synergic Control of Computer Manipulators", Proceedings of IV IFAC Symposium on Automatic Control in Space, Dubrovnik, Yugoslavia, Sept. 6-10, 1971, pp. 7.29-7.33.

[12]   Bekker, M.G., "Mechanics of Locomotion and Lunar Surface Vehicle Concepts", Paper 632K, Society of Automotive Engineers, Automotive Engineering Congress, Detroit, Michigan, U.S.A., Jan. 14-18, 1963.

[13]   McGhee, R.B., "Some Finite Aspects of Legged Locomotion", Mathematical Biosciences, Vol. 2, No. 1/2, pp. 67-84, February 1968.

[14]   McGhee, R.B., and Frank, A.A., "On the Stability Properties of Quadruped Creeping Gaits", Mathematical Biosciences, Vol. 3, No. 3/4, pp. 331-351, October, 1968.

[15]   McGhee, R.B., and Jain, A.K., "Some Properties of Regularly Realizable Gait Matrices", Mathematical Biosciences, Vol. 13, No. 1/2, pp. 179-193, February 1972.

[16]   Frank, A.A., and McGhee, R.B., "Some Considerations Relating to the Design of Autopilots for Legged Vehicles", Journal of Terramechanics, Vol. 6, No. 1, pp. 23-25, March 1969.

[17]   Okhotsimski, D.E., and Platonov, A.K., "Digital Computer Simulation of the Legged Locomotion System", Proceedings of IV IFAC Symposium on Automatic Control in Space, Dubrovnik, Yugoslavia, September 6-10, 1971, pp. 1.21-1.22.

[18] Pai, A.L., "Stability and Control of Legged Locomotion Systems", Ph.D. dissertation, The Ohio State University, Columbus, Ohio, U.S.A., 1971.

[19] Chen, D., Hemami, H., MacLean, I.C., and McGhee, R.B., "An Electrically Controlled Knee Joint for Long Leg Braces", Proceedings of Fourth International Symposium on External Control of Human Extremities, Dubrovnik, Yugoslavia, Aug. 28-Sept. 2, 1972.

[20] Orin, D.E., "A Simulation Study of a Computer-Assisted Manual Control System for Legged Vehicles", M.S. thesis, The Ohio State University, Columbus, Ohio, U.S.A., December, 1972.

[21] Muybridge, E., "Animals in Motion" Dover Publications, Inc., New York, N.Y., 1957 (first published in 1899).

[22] Gubina, F., "Stability and Dynamic Control of Certain Types of Biped Locomotion," Proceedings of Fourth International Symposium on External Control of Human Extremities, Dubrovnik, Yugoslavia, Aug. 28-Sept. 2, 1972.

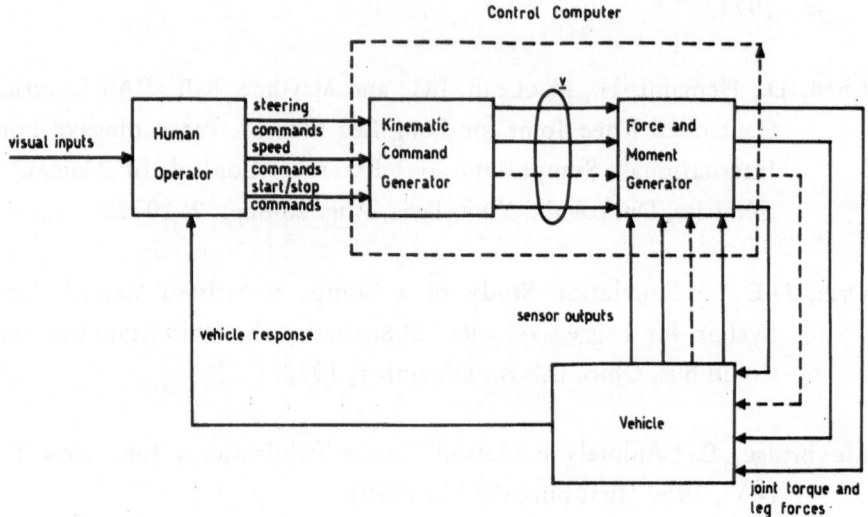

fig. 1 Block diagram of a legged vehicle supervisory control system

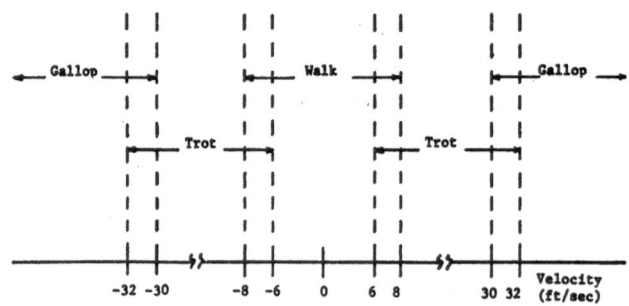

fig. 2 Automatic gait selection

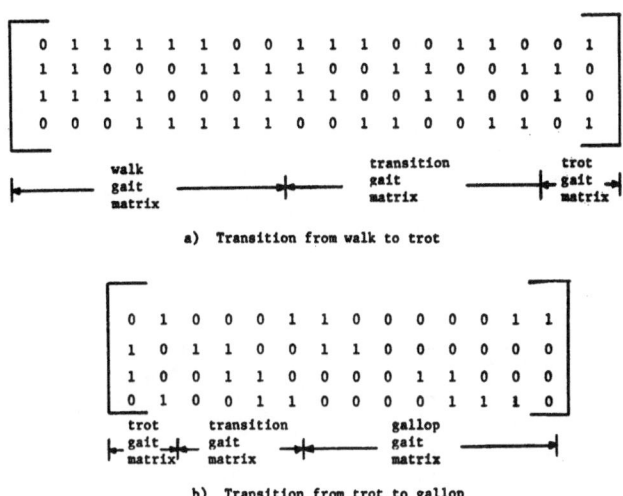

a) Transition from walk to trot

b) Transition from trot to gallop

fig. 3 Gait transition

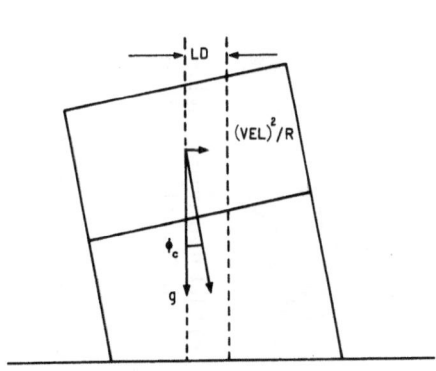

fig. 4 Rear view of a quadruped vehicle in a coordinated turn

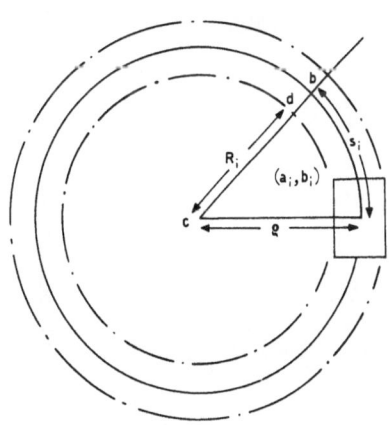

fig. 5 Top view of the quadruped vehicle moving along a circular path. The anticipated initial foot positions for front legs are shown with dots.

a) operator exercising joystick control

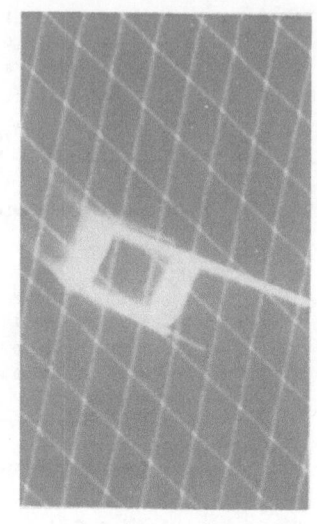

b) CRT display showing quadruped in halt state

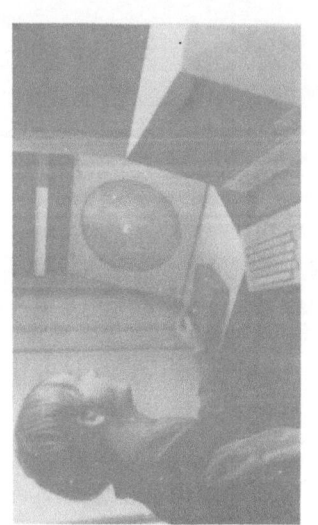

c) CRT display showing low speed turn in walk gait. Note: Display shows only weight-bearing legs.

d) CRT display showing high speed banked turn in trot gait.

fig. 6 Operator control station for simulated robot control system

A STUDY ON STABILITY OF BIPEDAL LOCOMOTION

Tadashi YAMASHITA, Associate Professor,
Hiroshi YAMADA, Research Fellow,
Kyushu Institute of Technology,
Department of Control Engineering,
Tobata, Kitakyushu, Japan

(*)

Summary

The theoretical study on bipedal locomotion is challenging because it may contribute to the understanding of human walking and to design of lower limb prosthesis and locomotion machine. Especially, the dynamic stabilization is one of the most important problems for the applications of theoretical results.

A control method to achieve a dynamically stable bipedal locomotion is presented. The machine is assumed to have a rigid body and two massless legs. The motion of the machine is analyzed by taking into consideration the frictional force between the foot and the ground. The proposed method can generate the similar walking patterns to human walking on the ground.

(*) All figures quoted in the text are at the end of the lecture.

1. Introduction

The theoretical study on bipedal locomotion is challenging because it may contribute to the understanding of human walking and to the design of lower limb prosthesis and locomotion machine. Especially, the dynamic stabilization is one of the most important problems for the applications of theoretical results. The purpose of this paper is to propose a simple control strategy for this problem.

The mathematical models used so far for the dynamic analyses of bipedal locomotions can be classified into two main classes. The first is concerned with a multiple-body mechanical model and analyzes its motion in a sinlge plane [1 - 4]. The second class considers a single-body mechanical model moving in a space [5 - 11]. As to the physical conditions for the locomotion system, the friction effect between the foot and the ground has been rarely analyzed [1-11]. The frictional force, however, greatly affects the dynamic stability of the bipedal locomotion system. For example, we know well that a walking on the ice is very difficult.

To get a clear idea about effective controls for stabilization of the system in a real situation, the influence of frictional force on the dynamic stability will be analyzed in a linear model derived by simplifying the other factors. The only fundamental concept will be considered here because this paper does not intend to design a specific machine. A simple control strategy which can control both the posture and the motion of a locomotion machine will be proposed based on the dynamic characteristics of the system. Finally, the walking patterns generated by the proposed strategy will be compared with human walking on the level ground.

2. Model of bipedal locomotion system

An idealized machine is used in this paper to study the dynamic stability of a bipedal locomotion. This model consists of a rigid body and two massless legs. Since a body of bipedal locomotion system is presumed to move in three dimensional space, the body of this system is assumed to have a mass and three moments of inertia about the principal axes. The legs of negligible mass in comparison to the mass of the body are attached to the body at the hips by ball joints.

The following assumptions are made to derive the linearized mathematical model for the system.

(1) All variables and their change rates which respect to time are small.

(2) The terms higher than the second order of state variables will be neglected, say $\sin \alpha = \alpha$, $\cos \alpha = 1$, $u \times v = 0$, etc., by the first order

approximation.

(3) The length of the leg ℓ is long enough to assume the magnitude of the term $1/\ell$ to be the same order of the magnitude of the angle variable.

(4) The distance of the hip from the center of gravity of the body is small compared with the length of the leg.

(5) The foot of the leg is kept in a supporting phase, that is a stance phase, at a certain point on the ground by static friction between the foot and the ground.

3. Linearized mathematical model

Fig. 1 shows right-handed rectangular coordinate systems and main symbols used to formulate the mathematical model,
where

X_E, Y_E, Z_E = earth-fixed coordinates, X_E; the direction of walking, Z_E; the direction of gravity (positve downward),

$x_1, y_1, z_1 (x^1, y^1, z^1)$ = auxiliary coordinates fixed at the center (hip) of the body parallel to the $X_E Y_E Z_E$ (xyz) coordinates,

x, y, z = moving body coordinates defined by successive rotations through the Euler angles ψ, θ, and ϕ of the $x_1\,y_1\,z_1$ coordinates,

x^ℓ, y^ℓ, z^ℓ = moving leg coordinates defined by successive rotations through the Euler angles α and β of the $x^1 y^1 z^1$ coordinates, z^ℓ; the direction of the leg,

u, v, w = components of the translational velocity of the center of gravity of the body expressed in earth coordinates,

p, q, r = components of the rotational velocity of the body expressed in body coordinates,

h_y, h_z = hip positions expressed in body coordinates.

X_F, Y_F, Z_F = foot positions expressed in earth-fixed coordinates,

$f_\ell (m_\alpha, m_\beta, m_\gamma)$ = control force (moments) defined in leg coordinates.

The following linearized equations are derived by the first order approximation from the accurate nonlinear equations expressing the dynamics of the body [5].

$$\dot{u} = -(\ddot{\alpha} + \ddot{\theta})\ell + \ddot{\psi}h_y - \ddot{\theta}h_z = f_x/m \tag{1}$$

$$\text{(2)} \qquad \dot{v} = (\dot{\beta} + \dot{\phi})\ell + \ddot{\phi}h_z = f_y/m$$

$$\text{(3)} \qquad \dot{w} = -\ddot{\phi}h_y = f_z/m + g$$

$$\text{(4)} \qquad \dot{p} = \ddot{\phi} = m_x/I_{xx}$$

$$\text{(5)} \qquad \dot{q} = \ddot{\theta} = m_y/I_{yy}$$

$$\text{(6)} \qquad \dot{r} = \ddot{\psi} = m_z/I_{zz}$$

where

$$m = \text{mass of the body,}$$

I_{xx}, I_{yy}, I_{zz} = moments of inertia about the x, y, and z axes,

g = gravitational acceleration,

f_x, f_y, f_z = components of the force acting on the body at the hip expressed in earth-fixed coordinates

m_x, m_y, m_z = components of the moment acting on the body about its center of gravity expressed in body coordinates.

In deriving this set of equations, it has been assumed that body coordinate axes are principal axes for the system and the derivative of leg length with respect to time is zero in the supporting phase.

The body force and moment vectors in the right leg supporting phase are related to the control force and moments defined in the leg coordinates by the following equations.

$$\text{(7)} \qquad \begin{bmatrix} f_x \\ f_y \\ f_z \end{bmatrix} = T_2\, T_1 \begin{bmatrix} -\,m_\alpha/\ell \\ m_\beta/\ell \\ -\,f_\ell \end{bmatrix}$$

$$\text{(8)} \qquad \begin{bmatrix} m_x \\ m_y \\ m_z \end{bmatrix} = \begin{bmatrix} 0 \\ h_y \\ h_z \end{bmatrix} \times T_1 \begin{bmatrix} -\,m_\alpha/\ell \\ m_\beta/\ell \\ -\,f_\ell \end{bmatrix} - T_1 \begin{bmatrix} m_\beta \\ m_\alpha \\ m_\gamma \end{bmatrix}$$

where

$$T_1 = \begin{bmatrix} 1 & 0 & \alpha \\ 0 & 1 & -\beta \\ -\alpha & \beta & 1 \end{bmatrix} \tag{9}$$

$$T_2 = \begin{bmatrix} 1-\psi & \theta \\ \psi & 1-\phi \\ -\theta & \phi & 1 \end{bmatrix} \tag{10}$$

The physical conditions which make sure of keeping the foot at a certain position on the ground without slippage are expressed mathematically in the following relations [11],

$$f_x^2 + f_y^2 = \{m_\alpha/\ell + (\alpha + \theta) f_\ell\}^2 + \{m_\beta/\ell + (\beta + \phi) f_\ell\}^2 \leq (\eta f_\ell)^2 \tag{11}$$

$$|m_\gamma/a| \leq \eta f_\ell \tag{12}$$

where
η = coefficient of static friction,
a = moment arm related to the foot size.

Let us define an effective coefficient of friction η_e to express the admissible regions of control moments m_α and m_β separately. A suitable definition for η_e is,

$$\eta_e = \eta/\sqrt{2} \tag{13}$$

then we get the following relations.

$$\left| m_\alpha/(\ell f_\ell) + \alpha + \theta \right| \leq \eta_e \tag{14}$$

$$\left| m_\beta/(\ell f_\ell) + \beta + \phi \right| \leq \eta_e \tag{15}$$

4. Stable region in phase plane

The reachable space from a point in the state space can be defined because the admissible control moments are limited as stated above. Using the concept of reachable space, the stable subspace could be found out.

A suitable definition for the stable subspace may be something like this; a stable subspace is the subspace in which there exists at least one state trajectory that can be transferred to the origin of the phase space by a proper control of force and moments without changing the supporting leg. The condition of not changing the supporting leg would be reasonable, since in this paper we have assumed the leg of massless type. The massless leg can be placed instantaneously at any point desired to stabilize the body motion.

Let us consider the stable region for the translational motions in the $X_E Z_E$ and $Y_E Z_E$ planes. From the pair of Eq. (7) and either Eq. (1) or Eq. (2), the following normalized equation is derived.

(16)
$$\overset{\circ\circ}{Y} - Y = M$$

where $Y = \alpha + \theta \ (\text{or} = \beta + \phi)$,

$\overset{\circ\circ}{Y}$ = the second order derivative of Y with respect to the nondimensional time $t\sqrt{f_\varrho/m\ell}$,

$$M = m_\alpha/\ell f_\varrho \, (\text{or} = m_\beta/\ell f_\varrho).$$

In deriving Eq. (16), the smaller terms, for example ψh_y , ψ/ℓ etc., have been substituted by zeros.

Considering the limitation of the moment M

(17)
$$| \, M + Y \, | \leqq \eta_e$$

the stable region for the Y motion has been found out as shown in Fig. 2. In the stable region we can find a control which can transfer a state to the origin of the phase plane. Although from the unstable region any state cannot be transferred to the origin of the plane by only manipulating the moments, the trajectory can be transferred to the stable region by changing the supporting leg and choosing a proper step length. The symbol USR-K in Fig. 2 shows an unstable region in the one-leg supporting phase, but from which state trajectories can be transferred to the stable region by changing of the supporting leg at least k-times.

Fig. 2 implies an effectiveness of changing a supporting leg and choosing a proper step length for the stabilization of the system.

5. Simple control strategy

We have to control the posture of the body as well as the propulsion. Let us consider a very simple control strategy which can satisfy this requirement and is easily analyzed.

Let us assume the following locomotion mode.

(1) No rotations of the body about its center of gravity.

(2) Always only one leg is in the supporting phase, that is, the supporting leg is changed instantaneously.

The values of control force and moments for this mode are determined for the right leg supporting phase as follows.

$$f_\ell = mg \tag{18}$$

$$m_\alpha = -\alpha h_z mg/(1 + h_z/\ell) \tag{19}$$

$$m_\beta = -(h_y + \beta h_z)mg/(1 + h_z/\ell) \tag{20}$$

$$m_\gamma = 0 \tag{21}$$

In the left leg supporting phase, the term h_y should be replaced by $-h_y$ because of the structural symmetry of the body.

Substituting Eqs. (18) through (21), the equations of motion are reduced from Eqs. (1) through (6) to the following simplified forms.

$$\overset{\circ\circ}{\alpha} - \alpha = 0 \tag{22}$$

$$\overset{\circ\circ}{\beta} - \beta = -h_y/\ell \tag{23}$$

In this set of equations, the derivatives have been calculated with respect to the nondimensional time $t\sqrt{g/(\ell+h_z)}$. The initial conditions for the ϕ, θ, and ψ have been assumed to be zeros.

Also substituting Eqs. (18) through (20) into Eqs. (14) and (15), the admissible regions for angles α and β are derived as follows.

$$|\alpha| \leq (1 + h_z/\ell)\eta_e \tag{24}$$

$$|\beta - h_y/\ell| \leq (1 + h_z/\ell)\eta_e \tag{25}$$

The admissible region for the angle β is biased. Therefore, if the coefficient of friction is smaller than the certain value, the foot has to be placed inward relative to the hip.

By choosing the step length within the angle limitations expressed in Eqs. (24) and (25), the stable locomotion can be realized.

6. Walking pattern by proposed control

The proposed control has been rather intuitively derived by using the simplified linear model. However, our approach can be justified to some extent. First of all, as to the assumption of no rotations of the body about its center of gravity, the rotations of the body are much smaller than the leg angles in a human walking [12]. Secondly, as to the linearized model, a linear approach is the easiest tool and is also always beneficial in an early stage of any research. In point of fact, whether the nonlinear terms play essential roles in a bipedal locomotion or not has not been completely examined yet to the authors' knowledge.

Therefore, comparing the walking pattern generated by the proposed control with that of a hyman walking might be interesting. A generated walking is shown in Fig. 3 in phase planes. This walking starts from a quiet standing pose. Although the first step is taken backward to the direction of walking, the body moves toward the desired direction. Since the third step, the system gets to a steady motion.

In Fig. 4, the motion of the body center in the $X_E Y_E$ plane for the trajectory shown in Fig. 3 is compared with that measured in a slow walking of a man [13].

In Fig. 5, the calculated reaction forces from the ground are compared with the measured ones [14].

Consequently, the walking generated by the proposed control can demonstrate the similar pattern to the pattern of a human walking as far as the level walking is concerned.

7. Conclusion

The dynamic stability of a bipedal locomotion system with massless legs was studied in a simplified model by the first order approximation. A simple control scheme for hip moments which can maintain the body in an upright posture, together with the step length control and changing the supporting leg, has been successfully applied to generate a stable walking. The generated walking has demonstrated the similar pattern to a human walking on the level ground.

The step length control and the changing of the supporting leg are indispensable means for stabilization of the system because applicable control moments are limited by the friction between the foot and the ground.

Acknowledgement

One of the authors, Yamashita, whishes to express his gratitude to professor A.A. Frank for his continuous discussion during the research period spent at the Department of Electrical Engineering, of the University of Wisconsin.

REFERENCES

[1]   VUKOBRATOVIĆ M., JURIČIĆ D., Contribution to the synthesis of biped gait, Trans. of IEEE, BME-16, 1, 1969, pp. 1-6.

[2]   VUKOBRATOVIĆ M., JURIČIĆ D., FRANK A.A., On the control and stability of one class of biped locomotion systems, Trans. of ASME, D, 92, 1970 pp. 328-332.

[3]   CHOW C.K., JACOBSON D.H., Studies of human locomotion via optimal programming, Math. Biosci., 10, 1971, pp. 239-306.

[4]   VUKOBRATOVIĆ M., HRISTIĆ D., ĆIRIĆ V., ZEČEVIĆ M., Analysis of energy demand distribution with anthropomorphic systems, Presented at the fourth int. sym. on External Control of Human Extremities, Dubrovnik, Yugoslavia, 1972.

[5]   FRANK A.A., Automatic control systems for legged locomotion machines, USCEE Report 273, Univ. of Southern California, Los Angeles, 1968.

[6]   FRANK A.A., An approach to the dynamic analysis and synthesis of bipedal locomotion machines, Med. and Bio. Engg., 8, 1970, pp. 465-476.

[7]   WITT D.C., A feasibility study on automatically controlled powered lower-limb prostheses, Univ. of Oxford Science Report 1119/70, 1970.

[8]   PAI A.L., Stability and control of legged locomotion systems, Ph. D. Dissertation, Ohio State Univ., Columbus, 1971.

[9]   TOWNSEND M.A., SEIREG A., The synthesis of bipedal locomotion, J. of Biomech., 5, 1972, pp. 71-83.

[10] GUBINA F., Stability and dynamic control of certain types of biped locomotion, Presented at the fourth int. sym. on External Control of Human Extremities, Dubrovnik, Yugoslavia, 1972.

[11] YAMASHITA T., FRANK A.A., A study of controllability of body motion in a biped by a linearized model, to appear.

[12] MURRAY M.P. DROUGHT A.B., KORY R.C., Walking patterns of normal men, J. of Bone and Joint Surgery, 46-A, 2, 1964, pp. 335-360.

[13] KATOH R., Patterns of vertical resultant force in standing and walking, M.S. Thesis, Kyushu Inst. of Tech., 1973 (in Japanese).

[14] CUNNINGHAM D.M., Components of floor reactions during walking, Univ. of California, Berkeley, 1958.

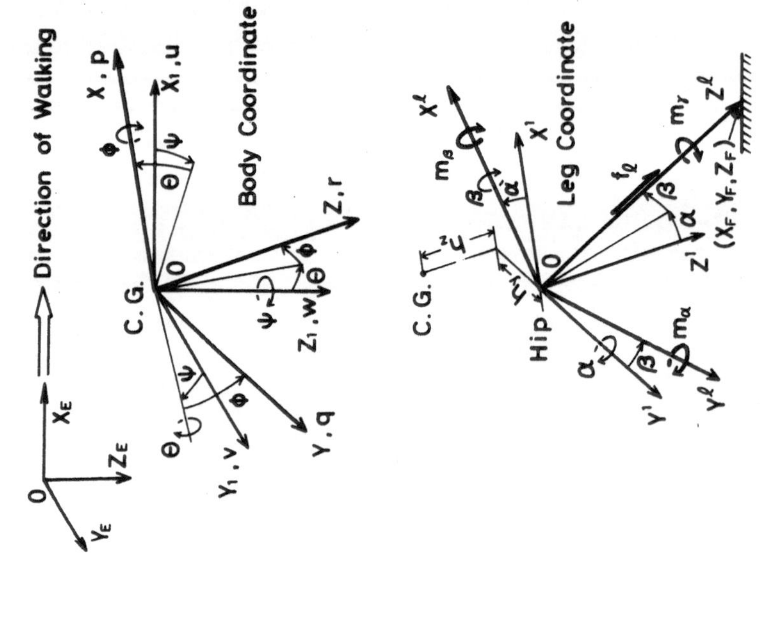

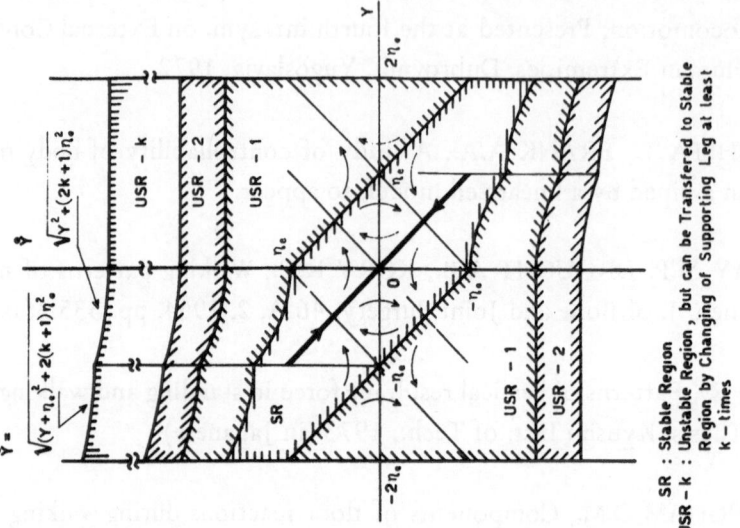

Fig. 2 Stable region expressed in phase plane

SR : Stable Region
USR - k : Unstable Region, but can be Transfered to Stable
         Region by Changing of Supporting Leg at least
         k - times

Fig. 1 Coordinate systems and main symbols

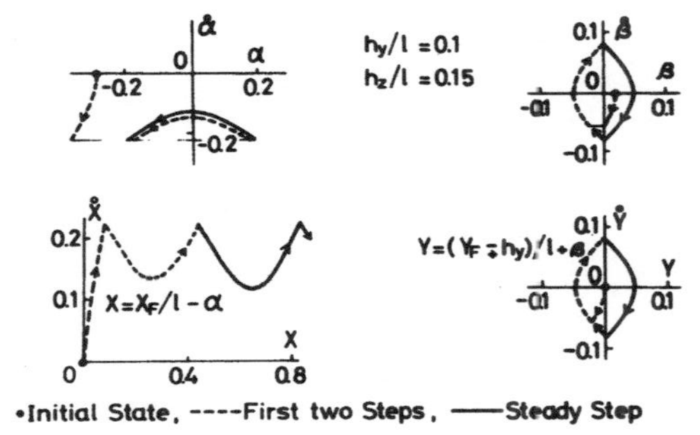

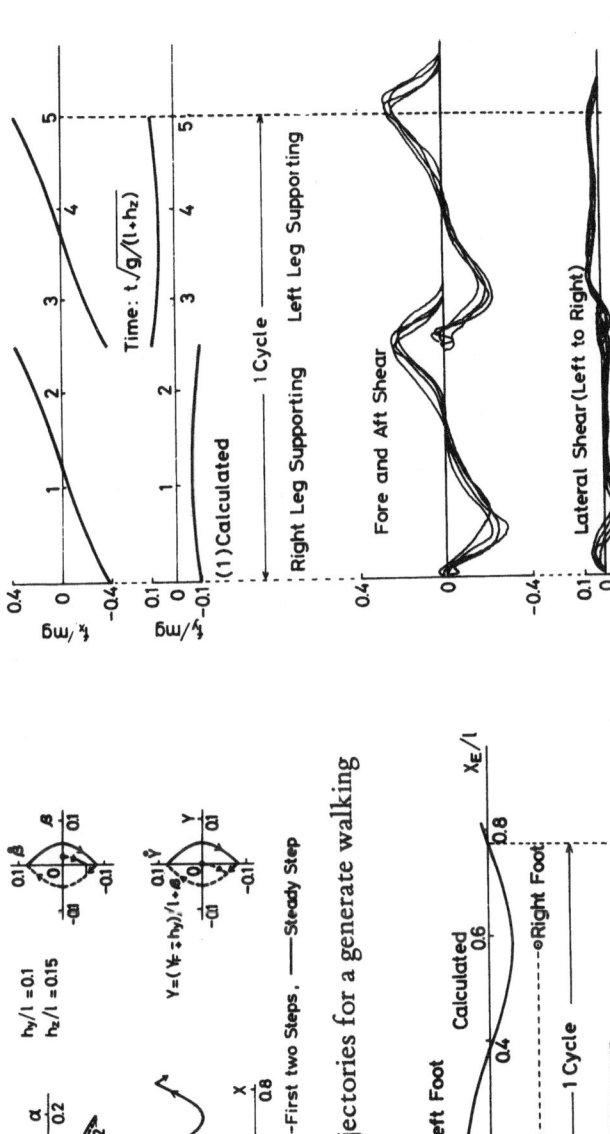

Fig. 5 Reaction forces in steady motion

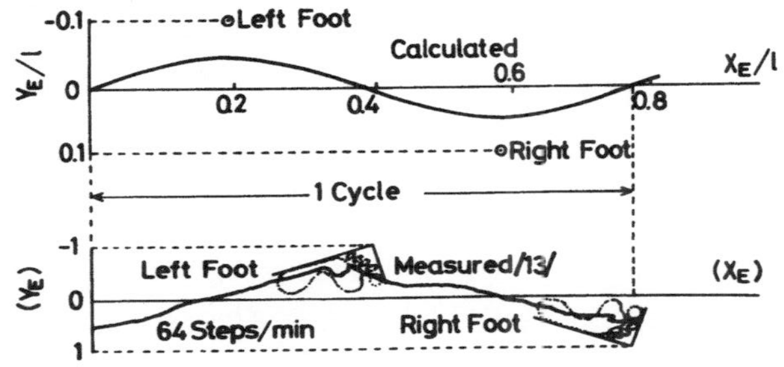

Fig. 4 Steady motion of body center in X_E, Y_E plane

Fig. 3 Phase trajectories for a generate walking

2. KINEMATICS AND DYNAMICS

A GENERALIZED METHOD FOR SOLVING A GROUP OF PROBLEMS REFERRING TO THE DYNAMICS OF MANIPULATORS WITH AN ELECTROMECHANICAL SERVO DRIVE

Ivan I. ARTOBOLEVSKII, Academician,
Academy of Sciences of the USSR, Moscow, USSR

Alexander G. OVAKIMOV, Assistant Professor,
Moscow Aviation Institute, Moscow, USSR

(*)

Summary

To describe the drive configuration, transmission systems and their dynamic parameters, a matrix group is used. This makes it possible to carry out dynamic calculations for manipulators with different drive layout by means of common calculation formulae.

The solution of a group of dynamic problems is given. The displacement of the actuator mechanism, according to the present motion of the control lever, is described. The influence of the system dynamics on the error, when transmitting the clamping action of the terminal device, and on the perception by the operator of the load applied to the actuator mechanism is discussed.

The displacement of the actuator machanism under the action of static loads is also defined.

(*) All figures quoted in the text are at the end of the lecture.

The control and actuator mechanisms of a reproducing manipulator (1) are mechanical arms of identical scheme and design. The servo drive is one type of connection combining these two mechanisms into one system. This connection manifests itself in a definite series of efforts that are transferred from the control to the actuator mechanism and backwards when the position and speed of these two manipulator mechanisms are not identical.

The terminal device, 6, of the mechanical arm (for example, fig. 1.a) has, as a rule, six degrees of freedom. Notwithstanding the layout of the kinematic chain, its position will be defined by six generalized coordinates q_1, ..., q_6; i.e. parameters determining the relative link position in the chain kinematic pairs. It is convenient to number the links $\nu = 1, ..., 6$ and the coordinates $q_1, ..., q_6$ beginning from the frame 0.

The seventh degree of freedom of the mechanical arm is defined by the generalized coordinate q_7. Its change affects the displacement of the terminal device gripper h with respect to its body. In the frequently used terminal device design (fig. 1.b) it is convenient to take as q_7 the angle of rotation φ_{76}, of one of the cranks OK or OK', which is connected kinematically to h.

The drive structure depends on the place where the motors are mounted. Three kinds of drives are available.

In the first type of drive all motors of the mechanical arm are fixed on a frame. A gear or a belt-cable transmission system leading to each motor has a differential scheme because of motion superposition. The idea of such a scheme may be derived from fig. 1.b, which represents a gear drive system running from the last three generalized coordinates

$$q_k = \varphi_{k,k-1}$$

(k = 5,6,7) to coaxial wheels 5", 6", 7" and to relative motors. The angle of rotation of these wheels is influenced by the variance of the previous chain coordinates - angles φ_{10}, φ_{21}, φ_{32} and φ_{43} (fig. 1,a).

In the second type of drive the motors are mounted on links 0,1, ..., 6 of the chain and each of them changes only one of the generalized coordinates q_1, q_2, ..., q_7. The drive of the third type is a combined system. In this case some motors are mounted on the base and the other part - on moving links.

The following solution of static and dynamic problems of manipulators is based on estimated formulae common for all three drive types. The structure of the, drive, gear system, as well as their inertia, is reflected in the initial data in the group of special matrices.

The solution of static and dynamic problems of manipulators is connected with the analysis of both its mechanisms - the control and the activator ones. In our case we impose certain restrictions on a part of $7 + 7 = 14$ variable parameters of such a system.

Let q_1, ..., q_7 - be variable parameters defining the position of the actuator mechanism and grippers on its terminal device and q'_1, ..., q'_7 - be the same variables on the control mechanism. Other values, referring to the control mechanism and which may have a different magnitude, will be also differentiated by means of the "prime".

Let us suppose that the operator displaces the control level (i.e. displaces the terminal of the control mechanism) according to a law prescribed beforehand. Then the variables q'_1, ..., q'_6 are determined functions of time t. Let us also assume that a solid body - the manipulated object - is mounted in the terminal device. This gives the condition $q_7 = $ const, which is the seventh restriction condition.

The system under investigation has seven degrees of freedom to which correspond generalized coordinates q_1, ..., q_6 and q'_7. The first six values refer to the actuator mechanisms and the last one - to the control mechanism.

The following problems are to be solved:

1. To determine the position errors of the statically loaded activator mechanism.

2. To determine the actuator mechanism movement according to the preset displacement of the control lever.

3. To determine the dynamic and static errors when the holding effort is transmitted to the manipulated object.

4. To determine the operator force action and the dynamic errors of sensitivity when reproducing the preset control lever displacement.

The first of these problems arises when evaluating the manipulator precision. The presence of a load on the actuator mechanism calls forth definite displacements in the servo drive, which grasps and balances this load. As a result of these displacements, the actuator mechanism takes a position different from that of the control mechanism. The problem is to determine six values $\Delta^\circ_k = q_k - q'_k$ ($k = 1$, ..., 6) - the statical errors of the actuator mechanism position. Using these data, it is possible to determine the true position of the manipulated object.

The following three problems answer the questions on the true displacement of the actuator mechanism and on the influence of the system dynamics on the transmitted and accepted efforts.

The solution of the problems given above requires the use of a number of kinematic values defining the position and displacement of the links of the control and actuator mechanisms.

The determination of efforts (moments) arising in servo drives is also a means that helps to solve the problems. This becomes complicated owing to the fact that when the misalignment Θ_i and the misalignment velocities $\dot{\Theta}_i$ in the drive number i = 1, ..., 7 are determined, one should consider the possibility of the above mentioned displacement superposition.

To describe mathematically the structure of the mechanical arm drive let us consider the Jacobi matrix (2)

$$(1) \qquad \Psi = \frac{D\psi}{Dq} = \begin{Vmatrix} \dfrac{\partial \psi_1}{\partial q_1} & \dfrac{\partial \psi_1}{\partial q_2} & \cdots & \dfrac{\partial \psi_1}{\partial q_7} \\[2mm] \dfrac{\partial \psi_2}{\partial q_1} & \dfrac{\partial \psi_2}{\partial q_2} & \cdots & \dfrac{\partial \psi_2}{\partial q_7} \\[2mm] \cdot & \cdot & \cdots & \cdot \\[2mm] \dfrac{\partial \psi_7}{\partial q_1} & \dfrac{\partial \psi_7}{\partial q_2} & \cdots & \dfrac{\partial \psi_7}{\partial q_7} \end{Vmatrix}$$

In its elements the quotient gear ratios $\partial \psi_i / \partial q_k$ (i, k = 1, ..., 7) for the angle $\psi_i = \psi_i (q_1, ..., q_7)$ of the motor rotor, number i, with respect to its body are given.

When the determining gear ratios $\partial \psi_i / \partial q_k$, we consider seven different one degree of freedom mechanisms that may be produced by the mechanical arm if only one of its generalized coordinates, each time a different one, is considered as variable.

For all drive types the matrix Ψ elements are constant values. With the appropriate choice of q_7 coordinate, the value $\partial \psi_7 / \partial q_7$ will not be an exception to the rule. For a drive of the second type Ψ has the form of a diagonal matrix.

The kinematic chain links of the mechanical arm are called carriers. Gear drive axes of rotation and in some cases drive elements (motors) are mounted on these carriers.

In the kinematics of moving elements number $u = p$, $p + 1$, ..., r on carrying member number ν, we are interested in their displacement with respect to the carrying member. Because of the displacement superposition, the relative angle of rotation α_u of the moving element u is the following function:

$$\alpha_u = \alpha_u(q_{\nu+1}, \; q_{\nu+2}, \ldots, q_7).$$

In order to describe the kinematics of the relative movement of the moving elements on the carrier ν, let us draw up a matrix A_ν from quotient gear ratios $a_{us} = \partial\alpha_u/\partial q_s$ ($s = \nu + 1, \nu + 2$, ..., 7) of those elements u, the inertia of which will be taken into account in equations of motion.

The gear ratios $\partial\alpha_u/\partial q_s$, the unit vectors $\bar{n}_u$ of axes of rotation of moving elements u, as well as their moments of inertia J_u with respect to axes of rotation, is used for determining vectors:

(2)
$$W_s^{(\nu)} = \sum_u \bar{n}_u J_u \frac{\partial\alpha_u}{\partial q_s} \qquad (s = \nu+1, \; \nu+2, \ldots, 7)$$

which are invariable in coordinate axes of the carrier member ν. These vectors are represented as columns in matrix W_ν which, together with the vector $J^{(\nu)}$ of moments of inertia $J_p^{(\nu)}$, ..., $J_r^{(\nu)}$ is connected with initial data used for calculations.

In the structure of the mechanical arm of reproducing manipulators there is one important feature - the kinematic chain, connecting the control lever (or terminal device) with the frame, is a structural group.

The displacement of the control lever, which is a driver link with six degrees of freedom, is prescribed with the help of a number of functions. Among them there is the law of coordinate variance of the chosen point H on axis x_6 (fig. 1.a), as well as functions, describing the rotary motion of the link about point H. In the calculation program (fig. 2) these functions are represented in the replaceable block 1.

According to the known displacement of the driving link, we determine by means of the kinematical analysis of the control mechanism driven link the variables q_1', ..., q_6' and their first and second time derivatives.

We consider the actuator mechanism (its carrier links as an open kinematic chain with six degrees of freedom. The aim of the kinematic analysis is to determine the angular velocities $\bar{\omega}_\nu$ and accelerations $\bar{\varepsilon}_\nu$ of its links ν, as well as accelerations $\bar{a}_{c_\nu}$ of points C_ν - centres of masses on these links.

The problem is solved by means of known values $\dot{q}_1, ..., \dot{q}_6$ and $\ddot{q}_1, ...,$ $\ddot{q}_6$ by using general formulae of the theorem on the velocity and acceleration addition. The position of link axes and the chain kinematic pairs is found beforehand according to the known values of generalized coordinates with the help of rotation matrics.

The problem of kinematic analysis comprises, as well, the determination of angle velocity analogues $\overline{\Omega}_k^{(\nu)}$ of links ν and velocity analogues $\partial \bar{r}_{c_\nu}/\partial q_k$ of points C_ν and some points of link 6. The kinematic sense of these vectors is clear from the following formulae:

$$\bar{\omega}_\nu = \sum_{k=1}^{6} \overline{\Omega}_k^{(\nu)} \dot{q}_k \;, \quad \bar{v}_{c_\nu} = \sum_{k=1}^{6} \frac{\partial \bar{r}_{c_\nu}}{\partial q_k} \dot{q}_k \;.$$

We have a group of formulae by means of which 3 x 6 matrices $\Omega^{(\nu)}$ and R_{c_ν} for each element ν and for given points are made up. Projections of vectors $\overline{\Omega}_1^{(\nu)}, ..., \overline{\Omega}_6^{(\nu)}$ and $\partial \bar{r}_{c_\nu}/\partial q_1, ..., \partial \bar{r}_{c_\nu}/\partial q_6$ are given in the matrix columns.

For the purpose of dynamic analysis it is advisable to represent the complex displacement of the actuator mechanism as a sum of two motions - the principal one and the additional one. The principal motion differs from the real one by the fact, that we suppose all the generalized accelerations $\ddot{q}_1, ..., \ddot{q}_6$ to be zero. On the contrary, in the additional motion all the generalized velocities $\dot{q}_1, ..., \dot{q}_6$ are equal to zero. The angle and line accelerations of the principal motion $\bar{\epsilon}_\nu^*$ and $\bar{a}_{c_\nu}^*$ will be required to calculate the parameters of the equations of motion.

In each drive of number i = 1, ..., 7 the actuator and control shafts are affected by moments M_i and $M_i' = -M_i : \mu$ connected by the proportionality coefficient μ. To calculate the moment

$$M_i = M (\Theta_i, \dot{\Theta}_i) \tag{3}$$

it is necessary to determine the misalignment $\Theta_i = \psi_i - \psi_i'$ and its time derivative.

The general solution of this problem is connected with the use of values $\Delta_k = q_k - q_k' (k = I, ..., 6), \Delta_7' = q_7' - q_7$ and their time derivatives. They are called position and velocity errors.

To determine the misalignment, the following formulae are available:

$$\Theta_i = \sum_{k=1}^{6} \frac{\partial \psi_i}{\partial q_k} \Delta_k \;, \quad (i = 1, ..., 6) \tag{4}$$

$$\Theta_7 = \sum_{k=1}^{6} \frac{\partial \psi_7}{\partial q_k} \Delta_k - \frac{\partial \psi_7}{\partial q_7} \Delta'_7 \tag{5}$$

To calculate values $\dot{\Theta}_1$, ..., $\dot{\Theta}_6$ and $\dot{\Theta}'_7$ let us use formulae which are derivatives of (4) and (5).

When determining static errors of position Δ°_1..., Δ°_6 and $(\Delta'_7)^\circ$, the clamping effort P'_7, produced by the operator for its transmission to the retaining body, must be taken into account. The effort P'_7 is a combination of two opposite forces of equal magnitude. If these forces (fig. 1.b) tend to increase the magnitude x'_7, it is recommended to consider $P'_7 > 0$.

When the manipulator is in static equilibrium, the effort P'_7 is balanced by the moment $M'_7 = -M_7 : \mu$, and the force of gravity $\bar{G}$ of the retained body - by moments M_1, ..., M_7 on the actuating shafts of the drive.

The necessary equations have been found with the help of the principle of possible displacements. One of these equations connects the effort P'_7 and the moment M_7,

$$M_7 \frac{\partial \psi_7}{\partial q_7} = \mu \, P'_7 \frac{\partial x'_7}{\partial q_1} \tag{6}$$

and for the lever design (or terminal device) of fig. 1.b it has the following form:

$$x_7 = 2(\tau \cos\varphi_{76} - a), \quad \varphi_{76} = q_7 . \tag{7}$$

The remaining six equations, linear with respect to unknown moments M_1, ..., M_7 are:

$$\sum_{i=1}^{6} M_i \frac{\partial \psi_i}{\partial q_k} = - \left(\bar{G} \cdot \frac{\partial \bar{r}}{\partial q_k} + M_7 \frac{\partial \psi_7}{\partial q_k} \right) \qquad (k = 1, \ldots, 6) \tag{8}$$

where $\partial \bar{r}/\partial q_k$ - the velocity analogue of the mass centre of the retained body.

With the help of equation systems (6), (8) and equations (3) - (5) we determine, by the iteration method (see fig. 2), the unknown moments, as well as the position errors Δ°_1, ..., Δ°_6 and $(\Delta^\circ_7)'$.

Let us continue with equations of motion. The mechanical arm is a holonomical system with stationary restrictions. The structures of Lagrange equations of motion of such a system has the following form (3):

(9) $I_{k1} \ddot{q}_1 + I_{k2} \ddot{q}_2 + \ldots + I_{k7} \ddot{q}_7 + F_k = Q_k$ $(k = 1, \ldots, 7)$

where

$$F_k = \sum_{s=1}^{7} \sum_{m=1}^{7} \left(\frac{\partial I_{ks}}{\partial q_m} - \frac{1}{2} \frac{\partial I_{sm}}{\partial q_k} \right) \dot{q}_s \dot{q}_m$$

and Q_k are the generalized forces. After transferring the values F_k to the right part of the equations, they are to be interpreted as generalized forces of inertia of the principal motion of the mechanism.

If written in a matrix form, equation (9) has the following form:

(10) $I\ddot{q} + F = Q$,

where $I - 7 \times 7$ is the matrix of coefficients $I_{ks} = I_{sk}$ and $\ddot{q}$, F and Q are vector-columns, consisting of seven synonymous values.

The determination of the vector-column Q of generalized forces is connected with the use of matrices Ψ, $\Omega^{(\nu)}$ and R_{c_ν} prepared beforehand.

The matrices I and F are determined as the sum of synonymous values $I^{(\nu)}$ and $F^{(\nu)}$, taking into account the inertia of the corresponding mechanism part. As such parts we take separate (loads, links), as well as whole groups of links - combinations of motors mounted on a frame or systems having the form of a carrying link with moving parts mounted on it.

For calculating inertia of motors and gear drives with fixed axes of rotation, we have the following formulae:

$$I^{(0)} = \Psi^T D\Psi \ , \ F^{(0)} = 0 \ ,$$

where D is a diagonal matrix of the inertia moments $J_1^{(o)} \ldots, J_7^{(o)}$ of rotors and reduced masses of drives and T - the transposition symbol.

The formulae for calculating the body inertia contain its mass, the inertia tensor in point C_ν and a number of kinematic values - the angle velocity $\overline{\omega}_\nu$, accelerations $\bar{a}^*_{c_\nu}$ and $\bar{\epsilon}^*_\nu$ in the principal system motion, as well as matrices R_{c_ν} and $\Omega^{(\nu)}$.

The formulae for calculating the inertia of the carrying link ν and moving elements mounted on it include, in addition, matrices A_ν, W_ν and $J^{(\nu)}$.

The parameters of seven equations for the mechanical arm motion are calculated in block 6 of the calculation program (fig. 2), according to general formulae. Two sets of these parameters, corresponding to the values of generalized coordinates and velocities of the actuator and control mechanisms, are used to solve the above mentioned problems of manipulator dynamics.

The solution of the problem of the manipulator motion is connected with the determination of generalized accelerations $\ddot{q}_1$, ..., $\ddot{q}_6$ and $\ddot{q}'_1$. With known values of generalized coordinates q_1, ..., q_6, q'_7 and velocities $\dot{q}_1$, ..., $\dot{q}_6$ $\dot{q}'_7$ these accelerations give the possibility to define new values of coordinates and velocities at the end of the small integration step Δt.

To solve the problem, we use six equations of motion of the actuator mechanism

$$\sum_{s=1}^{6} I_{ks} \ddot{q}_s = Q_k - F_k \quad (k = 1,\ldots,6) \tag{11}$$

and one equation of motion

$$I'_{77} \ddot{q}'_7 = Q'_7 - F'_7 - \sum_{s=1}^{6} I'_{7s} \ddot{q}_s \tag{12}$$

related to coordinate q'_7.

The errors of position Δ_1, ..., Δ_6, Δ'_7 and their derivatives are known values. Using these values, the efforts on the shafts of the servo drive are determined. Then the problem of generalized efforts of equations (11) and (12) is solved quite easily.

The effect of grippers on the retained solid body is a combination of two oppositely directed forces of equal value P_7. Then $-P_7$ designates the value of oppositely directed forces, that is the reaction of the body on the grippers. It is obvious, that the value $-P_7$ of these reaction forces is such, that in each moment of time the acceleration $\ddot{q}_7 = 0$.

Thereafter it is quite easy to formulate the principle of determining the effort P_7. It is necessary to compile an equation of motion, referred to the coordinate q_7. Into its left part $\&_7$ (T), besides the values $\ddot{q}_1$, ..., $\ddot{q}_6$ found earlier, one should substitute $\ddot{q}_7 = 0$, and in its right side, when determining the generalized force, it is necessary to take into account the reaction $-P_7$ on the grippers and the moment M_7 on the actuator shaft of the seventh servo drive.

As a result

(13)
$$P_7 = \left[M_7 \, \frac{\partial \psi_7}{\partial q_7} - \mathcal{E}_7 \, (T) \right] : \frac{\partial x_7}{\partial q_7}$$

where the gear ratio $\partial x_7 / \partial q_7$ is determined with the help of expressions, like those given in (7).

In this formulae the meaning of the function $\mathcal{E}_7 \, (T)$ must be summed as a known value. In this case, the coefficients I_{71}, ..., I_{77} and the value F_7 in formula (9) are defined according to the same values of generalized coordinates and velocities, as the system parameters (11).

The holding effort transmission error is the difference

(14)
$$\Delta P_7 = P_7 - \mu P_7'$$

between P_7 and the effort $\mu P_7'$, which must be transmitted to the body, taking into account the scale coefficient μ.

In statics the error in the process of transmitting the holding effort will be

(15)
$$\Delta P_7 = \mu \, P_7' \left(\frac{\partial x_7' / \partial q_7'}{\partial x_7 / \partial q_7} - 1 \right)$$

The cause of this error is quite obvious: the gear ratios $\partial x_7' / \partial q_7'$ and $\partial x_7 / \partial q_7$ in the control and actuator mechanisms are determined according to the argument values q_7' and q_7, that are not equal to each other.

The reproduction of the preset displacement of the control lever is equivalent to the superposition on the system of six restrictions by applying a combination of forces of definite value to the lever. Such system of forces must be determined by six scalar parameters. They may be represented eithe -by two forces $\bar{P}_1$ and $\bar{P}_2$ with preset points of application H_1 and H_2 or by force $\bar{P}_1$ and moment $\bar{M}_e$.

In the first six equations of motion for the mechanical arm of type (9) their left-hand parts $\mathcal{E}_k \, (T')$ should be assumed as known values.

Forces $\bar{P}_1$ and $\bar{P}_2$, which we are determining, should be introduced into generalized forces Q_1', ..., Q_6' with moments M_1', ..., M_7' on the control shafts of the servo drive. Then we have the following system of equations:

$$\frac{\partial \bar{r}_{H1}}{\partial q_k} \cdot \bar{P}_1 + \frac{\partial \bar{r}_{H2}}{\partial q_k} \cdot \bar{P}_2 = \mathcal{E}_k (T') - \sum_{i=1}^{7} M_i \frac{\partial \psi_i}{\partial q_k} \qquad (k = 1,\ldots,6) \quad (16)$$

By solving this system of equations, linear with respect to the projections of forces $\bar{P}_1$ and $\bar{P}_2$, it is possible to determine the operator effort.

Forces of opposite direction will define reactions $-\bar{P}_1$ and $-\bar{P}_2$ sensed by the operator. The problem of dynamic sensitivity errors is based on the comparison of these reactions and the load applied to the terminal device.

The described calculation is fully realized by means of the computing machine БЭ CM-6. Curves 1,2, ..., 6 (fig. 3) represent the laws of error position variance $\Delta_1 \Delta_2$...., Δ_6 of the actuator mechanism for one of the calculated cases.

REFERENCES

[1]   ARTOBOLEVSKII I.I., KOBRINSKII A.E., Robots (in Russian), Mashinovedenje, 5, 1970, pp. 3 -11.

[2]   OVAKIMOV A.G., ANSHIN S.S., The manipulator Jacobi matrix and its application to the determination of errors of state in statics (in Russian), Mashinovedenje, 2, 1972, pp. 34-39.

[3]   OVAKIMOV A.G., Equations of motion for space gear-lever mechanisms with some degrees of freedom (in Russian), Mashinovedenje, 6, 1970, pp. 28-34.

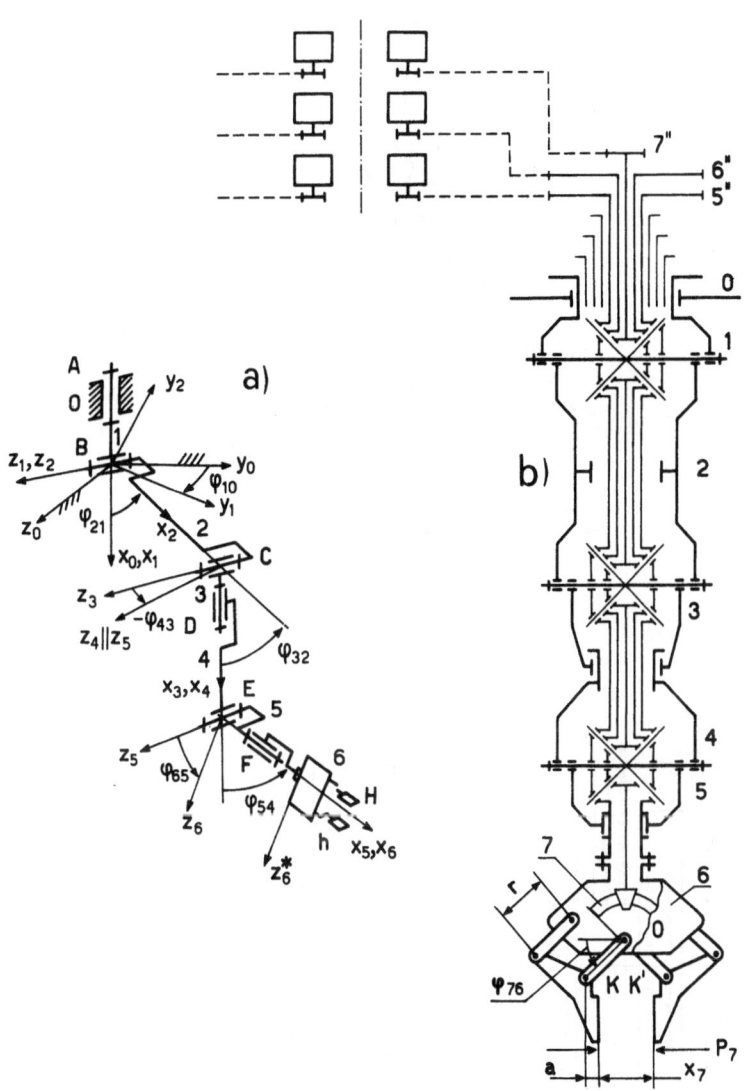

fig. 1

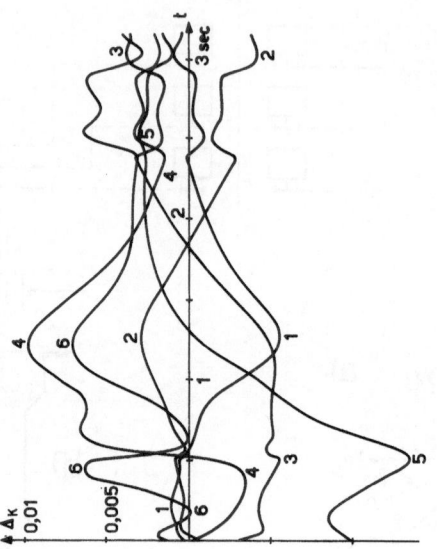

fig. 3

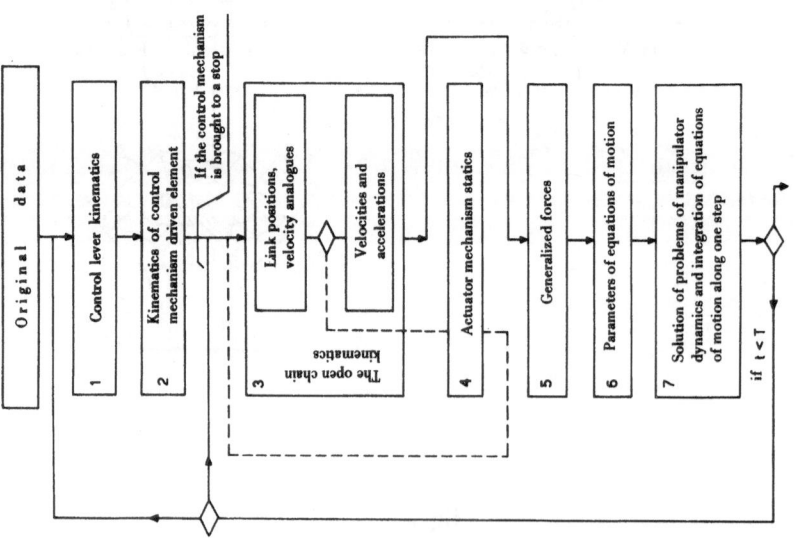

fig. 2 Block-diagram for the manipulator
dynamic calculation

A KINEMATICAL ALGORITHM AND DYNAMICAL POINT MASS SIMULATION APPLIED IN ROBOTS AND MANIPULATORS

Michael S. KOSTANTINOV, Professor,
Mechanical and Electrotechnical Institute,
Department of Mechanical & Technological Engineering,
Sofia, Bulgaria

Zanko I. ZANKOV, Assoc. Professor, Doctor,
Mechanical and Electrotechnical Institute
Department of Mechanical & Technological Engineering
Sofia, Bulgaria

(*)

Summary

A review which discusses current work on the development of manipulators and industrial robots is made. A mathematical algorithm for the kinematical investigation of manipulators and hand/arm systems of robots is conceived on the basis of a vector-complex operator. Thus, the relation between the movement of the object to be manipulated and the relative motion of the manipulator (hand/arm) linkages is clarified, wherein the distribution of the velocities in the end linkage (the hand) is determined. The dynamical response of the object is simulated by a system of permanent point masses. Finally, an example describing an automatic handling arrangement in hot forging industrial environment is discussed.

(*) All figures quoted in the text are at the end of the lecture.

1. Introduction

More than 2200 years ago, Aristotle wrote: "Then certainly, if each tool could act on received instructions or even guess the orders before performing the task ... and if the shuttle could weave and the zither could play by themselves, then no master would need any helping hand or servant." Indeed, was he fortune telling about a future automation age ?

Reference to the history of technological innovations shows that the advances in automation are impressive. However, the older automation age is dying and sophisticated robot development has been with us for approximately 25 years. The master's "helping hand" is a reality today. Within the next 10 years a considerable increase in the number of robots developed for industrial work can be expected. The future trend is to adapt the programmable handling machines to do routine tasks in the office, factory, and home. Interaction between artificial intelligence, programming and machine will take over tasks not suited for humans.

Forward thinkers suggest that no factory will be without blue collar robots. Dr. Driscoll [1] describes these robots as automatically programmable minirobots or lightweight factory manipulators with minicomputer or multi-processor control, multiple arms and hands, and interactive sensors. Minirobots will weight about 225 kg. and their payloads will be limited to about 20 kg. or less. He expects, furthermore, that many arms will resemble prostetic arms and that hands will have individual finger flexions. Dr. Driscoll predicts: "Blue-collar robots will evolve through programming synthesis of eye-hand coordination to highly context-dependent machine intelligence toward the end of the decade."

2. Manipulators and hand/arm systems

The earliest known antropomorphous devices in shape of artificial hands have been employed for prostheses in the old Egypt 2500 years before Christ and in Europe since about 300 years, namely as the hand of Götz von Berlichingen.

When one tries to design an artificial hand which is almost like a human hand there arise a lot of difficulties. A man has 27 degrees of freedom in his arm, hand and fingers. As an aside, it is interesting to note that inflatable fingers are currently available. Each finger is a hollow, half-bellows of glassfiber reinforced urethane. Under air-pressure, there is an unequal expansion and the fingers curl inwards to grasp.

All manual operated instruments and also the classical mechanical prostheses belong to the muscle-driven manipulators.

The activity in life hostile environments of nuclear energy, as well as for space and deep-sea research, lead to the development of teleoperated antropomorphous systems. A description of two types of handlers for radioactive materials is to be found in [2]. The anthropomorphous machine SYNTELMANN, an abbreviation of SYNchron-TEle-MANipulator, could be used in atomic energy and space research [3]. A man, called master, operates an exoskeleton, which is fixed on his arms and legs, synchronously to the corresponding limbs of the machine slave. A locomotion device is used to magnify the radius of action of a pair of arms. A series of deep-sea teleoperated manipulators was developed by the Institute of Oceanology, Academy of Sciences of the USSR [4].

The limitation of the human arm and hand is relevant for the function of the manipulators. To reach for, grasp and handle objects, a manipulator needs, theoretically, six degrees of freedom (degrees of manipulation) to have total spatial command (Fig. 1), three degrees to place the object in the desired spatial location and three to orient the object in the desired attitude. As a matter of fact, the configuration shown in the above figure is that of the Italian manipulator Mascot 1 [5]. The same articulated mechanical arrangement is used in the Japanese HI-T-HAND (Hitachi Tactile Controlled Hand) [6]. The entire system consists of a swivel arm, a flexible shoulder, a flexible elbow, a revolving forearm, a flexible wrist, and a revolving wrist.

More than six degrees of manipulation are required in order to increase the potentialities of the manipulator. These extra degrees are called the degrees of manoeuvrability [7] of the manipulator: it could reach around an obstacle (Fig. 2) or reorientate its geometric configuration, so that a complex class of motions is possible. Under these circumstances, the manipulator shown in Fig. 2 possesses two degrees of manoeuvrability. Generally speaking, it is normally less expensive and more satisfactory to use an arm with extra degree than to build-in special features for a hand tooling, or to move and align production equipment [8].

Fewer than six degrees might suffice if symmetry exists. As example, if an object is to be transferred from one horizontal plane to another, the pitch axis might be deleted. If the object has one-axis symmetry, such as a cylinder, a five axis manipulator will be able to place the cylinder in any desired position or attitude.

Every manipulator is provided with a gripper (Fig. 1) to hold and release the object being transported. In general, the grippers roughly simulate the movement of two fingers on a human hand. Thus, the manipulator is provided with a further degree of freedom. However, the manoeuvrability of the system is not improved by that extra degree of freedom.

As stated, the imitation of the human hand/arm is relevant for the function of manipulators. But this is not true for the physical effect thereof. For instance, when heavy weights (more than 10 kg.) are transported, human motion time is longer in all ranges of distance and the time increases in proportion to distance increases [9].

Another trend is to develop a nonhumanoid hand/arm system to perform the same work as the manipulator, as described and shown in Fig. 1. Its performance may not be manlike and it should have more flexibility than conventional types of special purpose manipulators. New ideal motion patterns could be conceived. We are entering in the field of industrial robots.

An industrial robot is defined as an automatically controlled handling device that can be reprogrammed for different work cycles. Industrial robots vary in complexity form pneumatically operated arms controlled by simple sequential control logic which can be set to carry out a predetermined sequence of movements, to sophisticated closed loop servo controlled machines in which the motions are stored in computer or magnetic storage systems and reproduced accurately on demand by the machine. Up to the present point in time, these are offered as (a) completely self contained devices comprising a hand/arm system with standard or tailor-made grippers, and (b) modularized systems comprising a manipulator and the application gripper. As a typical example, the arm/hand system (manipulator) of the majority of Versatrans in the field at the present (Fig. 3b) consists of one arm mounted horizontally on a vertical post, which in turn is mounted on a base permitting rotation of the post. At the end of the arm a wrist is provided which has two degrees of movement, sweep and rotate. The gripper is tailor-made to suit the customers application, but there is an increasing tendency to use the arm to manipulate a processing device, such as a spot-welding head or a paint spray gun.

Two further typical layouts of industrial robots are shown in Fig. 3 and there can be hybrid combination of these three configurations. Unimate (Fig. 3a) resembles a small off-white tank with a telescoping arm and a gyrating steel claw protruding from its turret where the gun should be. Gantrix (Fig. 3c) is an over-head gantry system that has versatility and can be fully programmed to give sequence changes.

3. Movements, working space, actuators

On analogy of the human hand, the manipulator or the hand/arm system of an industrial robot has a determined sphere of activity. Pick-and-place operations within this work area are performed by arm movements only, while a welding operation, for example, would be accomplished with the additional help of hand movements. The complex hand/arm motions require many degrees of freedom, these being closely related to the articulations of the mechanism.

If we pursue the arm and hand movements of a human, we will note that the arm motion capabilities present as many as five basic (primary) movements: sagittal, vertical, transverse, rotation, and swing. In addition to the basic arm motions the hand itself has three extra (secondary or wrist) movements: rotation, sweep, and swing. The primary movements perform the transfer of the object and define the location of the wrist as a point in the space, while the secondary movements accomplish the displacement of the object and define its oriented attitude.

From factual information obtained about the state and extend of robot development [10], [11] in a number of industrialized countries, we were able to compile Table 1 (see Appendix).

Besides the differences in playload and reach capabilities, many other variations exist among today's industrial robots. An important difference is the ability or lack of ability to stop movements at intermediate positions. The programming method fall in two classes: Point to Point (P to P) or Continuous Path (CP). The most simple robots are limited to operations over relatively fixed distances, and are capable of few primary and secondary movements. The highly sophisticated machines have several primary and secondary movements, have work envelopes in excess of 3 cu.m., and move in complicated paths.

We could say that primary movements determine the basic working space, while secondary movements define the additional working area. Considering the geometry of the hand/arm system, the basic working space is the space described by the point (B) of rotation of the wrist, see Figs. 3 and 4. The reference point (A) of the object, i.e. its geometric center coinciding with the origin of its own coordinate system (Fig. 4), describes the additional working area, in all cases being a sphere.

We would note that the rated working space (RS) is determined by the envelope of the additional working area (AA) in movement. It is clear that this envelope is equidistant with the boundary surface of the basic working space (BS).

Finally, the real work envelope is determined by the shape of the object. When the robot is working in the same environment as people, adequate protective barriers are mandatory around the real work envelope.

In Fig. 5 is shown the generation of the basic working space (BS) pertinent to the industrial robot Kaufeldt type A 13 (Sweden). Generally speaking, the form of the BS depends on the kinematical structure and joints of the robot. The structure is the skeleton of the robot and consists of links and joints. All antropomorphous manipulators (Fig. 1) are provided with swivel joints only, while the hand/arm systems of industrial robots (Fig. 3) could be provided with sliding pairs also. Thus, the radial traverse of Unimate is realized by one single sliding (telescoping) pair. Versatran has two sliding pairs: each one for the radial and vertical traverses. Three sliding pairs perform the transversal, sagittal and vertical movements of the Gantrix robot.

As shown in Fig. 3, the polar structure of Unimate determines a spherical BS, the cylindrical structure of Versatran defines a cylindrical BS, and the gantry structure of Gantrix identificates a parallelepipedon BS.

In view of their driving-power industrial robots hand/arm systems make a distinction as follow: (a) electromechanical systems, (b) hydraulic actuators, and (c) pneumatic actuators. Hydraulic and pneumatic actuators are able to produce high force and slow speed without intermediate gearing. Modern constructions of such actuators enable to transform the potential energy of the working medium into mechanical work under avoiding of frictional and sliding sealings.

In fact, the development of electromechanical systems is very advanced. However, because of the high rotational speed ratio of the electrical driving motors, they need gears with high speed ratio and have a poor efficiency. The by-pass could be surmounted in using ironless direct current disc-rotor motors with low moment of inertia connected to strain wave transmitters [12]. Describing a blue collar robot Dr. Driscoll [1] explains: "Harmonic drive sets with high torque and low speeds may be used to articulate elbow and shoulder joints." It should be noted that "Harmonic Drive" is a trademark for strain wave transmitters (gears). We have shown in Fig. 6 a manipulator provided with four harmonic drive sets (1, 4, 5, and 6) driven by disc-rotor motors (2).

4. Motions geometry and kinematics

Hand/arm systems (manipulators) are open kinematic chains, in which the relative motions are elementar: rotation and translation. Superposition of these elementar motions, in an appropriate manner, results in a complicated positioning of objects in the space (Fig. 7).

The vector complex operator

$$\left(\frac{\overline{\lambda}_s}{\overline{\mu}_s} \right) = \left(\frac{\overline{\lambda}_{s+1}}{\overline{\mu}_{s+1}} \right) e^{i\varphi s} \quad ; \quad \overline{\nu}_s = \overline{\nu}_{s+1} \tag{1}$$

could be advantageously employed for geometrical analysis and synthesis of relative and absolute motions in hand/arm systems. Here $\overline{\lambda}, \overline{\mu}, \overline{\nu}$ are unit vectors of spatial coordinate systems, having a common axis z and being turned to an angle φ_s.

In (1), the conditional term

$$\overline{\eta} = \overline{\lambda} + i\overline{\mu} = \left(\frac{\overline{\lambda}}{\overline{\mu}} \right)$$

expresses a complex vector.

The relative disposition of three adjacent links (S−1, S, S+1), pertinent to the structural composition shown in Fig. 8, could be described by the recurrence operator algorithm:

$$\begin{cases} \left(\dfrac{\overline{\lambda}_{s-1}}{\overline{\mu}_{s-1}} \right) = \left(\dfrac{\overline{\lambda}_s}{\overline{\mu}_s} \right) e^{i\varphi_s} \quad ; \quad \overline{\nu}_{s-1} = \overline{\nu}_s \\[3em] \left(\dfrac{\overline{\mu}_s}{\overline{\nu}_s} \right) = \left(\dfrac{\overline{\mu}_{s+1}}{\overline{\nu}_{s+1}} \right) e^{i\Psi s+1} ; \overline{\lambda}_s = \overline{\lambda}_{s+1} \end{cases} \tag{2}$$

In fact, the three-linked composition in Fig. 8, is a basic group of the structural scheme of an arbitrary complex hand/arm system. Furthermore, the reversibility of algorithm (2) is its advantage in carrying out the matrix transformations:

$$\begin{vmatrix} \overline{\lambda}_p \\ \overline{\mu}_p \\ \overline{\nu}_p \end{vmatrix} = \begin{vmatrix} a_{11} & a_{12} & a_{13} \\ a_{21} & a_{22} & a_{23} \\ a_{31} & a_{32} & a_{33} \end{vmatrix}_{pq} \begin{vmatrix} \overline{\lambda}_q \\ \overline{\mu}_q \\ \overline{\nu}_q \end{vmatrix} \tag{3}$$

and respectively

$$
(4) \qquad \begin{vmatrix} \overline{\lambda}_q \\ \overline{\mu}_q \\ \overline{\nu}_q \end{vmatrix} = \begin{vmatrix} a_{11} & a_{21} & a_{31} \\ a_{12} & a_{22} & a_{32} \\ a_{13} & a_{23} & a_{33} \end{vmatrix}_{pq} \begin{vmatrix} \overline{\lambda}_p \\ \overline{\mu}_p \\ \overline{\nu}_p \end{vmatrix}
$$

In the above matrix equations (3) and (4), the coefficients a_{mn}^{pq} of the transposed matrices are calculated with the help of algorithm (2) for a complete variation of the indexes p and q in the limits from zero to six. The expressions of these coefficients are given in Table 2 (see Appendix). Because of analogy and identity in some relative motions, Table 2 contains data presenting only these motions enclosed in rectangles, pertinent to the digital schemes shown in the table.

The position of point 0_k (Figs 8 and 9) is determined by the vector coordinate

$$
(5) \qquad \overline{\ell}_k = \overline{0_1 0_k} = \sum_{s=1}^{k} \overline{\lambda}_s h_s \qquad \begin{matrix} s = 1,3,5,\dots \\ k = 1,3,5,\dots \end{matrix}
$$

while its velocity is defined by the expression

$$
(6) \qquad \overline{V}_k = \dot{\overline{\ell}}_k = \sum_{s=1}^{k} (\dot{\overline{\lambda}}_s h_s + \overline{\lambda}_s \dot{h}_s)
$$

Since

$$
\overline{\lambda}_s = \overline{\lambda}_0 a_{11}^{0s} + \overline{\mu}_0 a_{21}^{0s} + \overline{\nu}_0 a_{31}^{0s}
$$

it follows for equation (6)

$$
(7) \qquad \begin{vmatrix} V_{kx} \\ V_{ky} \\ V_{kz} \end{vmatrix} = \sum_{s=1}^{k} \begin{vmatrix} h_s \dot{a}_{11} + \dot{h}_s a_{11} \\ h_s \dot{a}_{21} + \dot{h}_s a_{21} \\ h_s \dot{a}_{31} + \dot{h}_s a_{31} \end{vmatrix}
$$

The angle velocity of an arbitrary link n (an even number) and $n + 1 = k$, respectively, is determined by (see also Fig. 9)

$$\begin{cases} \bar{\omega}_{n,o} = \sum_{s=1}^{n-1} (\bar{\nu}_s \dot{\varphi}_s + \bar{\lambda}_{s+1} \dot{\psi}_{s+1}) \\ \bar{\omega}_{n+1,o} = \bar{\omega}_{n,o} + \bar{\nu}_{n+1} \quad \bar{\omega}_{n+1} = \bar{\omega}_{k,o} \end{cases} \qquad (8)$$

Points N on the instantaneous axis of helicoid motion of the link $k = n + 1$ move with a minimum velocity V_{min}^N defined by the scalar product equation

$$V_{min}^N \omega_{n+1,o} = V_{min}^N \omega_{k,o} = \bar{V}_k \cdot \bar{\omega}_{k,o} \qquad (9)$$

In fact, equations (6), (8), and (9) define unambiguously the velocity field of the link k.

Concluding, we should note that the recurrence operator algorithm (2) could be applied to all kinematical structures of hand/arm systems and manipulators. Thus, the algorithm is applicable both to the primary and the secondary movements (see Item 2 of this paper). The data given in Table 2 (Appendix) are sufficient for the identification of the primary movements pertinent to the most complex structures. For the secondary movements only Table 2(A) covers the requirements. We have investigated the geometry and kinematic (up to the range of velocities) of the primary movements in several industrial robots and manipulators. Table 3 (see Appendix) contains the results of our investigations.

5. Point mass models

The smoothness of motions and the velocity conditions with regard to the transfer and displacement time of an object manipulated by a hand/arm system, depend both on the speed-torque characteristic of the actuators and the dynamical reactions due to the accelerations caused by the non-steady-state motion conditions.

Dynamical modelling of spatial mechanisms [13], and more particularly by point concentration of masses [14], brings physical clearness and facilitates the calculus labour in determining the inertia forces of spatially moved solids, such as the links of a manipulator or hand/arm system and the object to be transferred. The method of modelling is based on the circumstance that a solid performing a tridimensional motion is dynamically equivalent to a discrete system of permanent heavy points [15]which are invariably attached to the solid.

In Fig. 10, X Y Z denotes a fixed coordinate system and x y z is a mobile coordinate system invariably attached to the spatially moved solid Q. One of the concentrated masses m_k is denoted by the letter K, m is the mass of the solid, and G is its point of gravity. The necessary and sufficient conditions for dynamical equivalence between the solid and the concentrated mass system is reduced to three postulates:

— The masses of the solid and the system of a number of p heavy points should be equal

(10)
$$\sum_{k=1}^{p} m_k = m$$

— The points of gravity of the solid and the heavy points system should coincide

(11)
$$\sum_{k=1}^{p} \bar{\rho}_k m_k = \int_{(Q)} \bar{\rho}\,dm$$

— The tensors of the inertial ellipsoides of the solid and the heavy points system should be equal

(12)
$$\sum_{k=1}^{p} \begin{Vmatrix} (x_k^2) & (x_k\,y_k) & (x_k\,z_k) \\ (y_k\,x_k) & (y_k^2) & (y_k\,z_k) \\ (z_k\,x_k) & (z_k\,y_k) & (z_k^2) \end{Vmatrix} m_k = \begin{Vmatrix} I_{xx} & I_{xy} & I_{xz} \\ I_{xy} & I_{yy} & I_{yz} \\ I_{xz} & I_{yz} & I_{zz} \end{Vmatrix}$$

Following components are introduced in the inertia tensor

(13)
$$I_{xx} = m\,i_{xx}^2 = \int_{(Q)} x^2 dm \qquad I_{xy} = \int_{(Q)} xy\,dm$$

Considering an infinite spatial motion of a solid, having an arbitrary shape and disposition of the masses, the system of equations (10), (11), and (12) permits four point mass concentrations as a minimum dynamical model. Additionally, following admissions could be done

$$x_1 = |x_2| \qquad z_1 = z_2 = 0 \qquad y_1 = y_2 = 0$$

$$m_1 = m_2 \ ; \quad m_3 = m_4$$

$$y_3 = |y_4| \qquad x_3 = x_4 = 0 \qquad z_3 = z_4 \quad (14)$$

It follows from (10), (11), and (14)

$$
\begin{cases}
\dfrac{m}{2}\dfrac{z_G^2}{i_{zz}^2} = m_3 = m_4 \qquad \dfrac{i_{yy}\,i_{zz}}{z_G} = y_3 = |y_4|\,\dfrac{i_{zz}^2}{z_G} = z_3 = z_4 \\[4mm]
\dfrac{m}{2}\left(1 - \dfrac{z_G^2}{i_{zz}^2}\right) = m_1 = m_2 \qquad \dfrac{i_{xx}\,i_{zz}}{\sqrt{i_{zz}^2 - z_G^2}} = x_1 = |x_2|
\end{cases}
\qquad (15)
$$

The system of concentrated masses according to (14) and (15), shown in Fig. 11, is recommended as a dynamical model for an object manipulated by a hand/arm system.

It will be noted that usually the links of the spatial mechanisms possess degenerated inertial ellipsoides. When the solid possesses one symmetry axis, the inertial ellipsoid degenerates in a figure of revolution — the spheroid. In this case, two of the inertia moments of the main diagonal of the tensor (12) are equal ($I_{xx} = I_{yy}$), consequently z is the symmetry axis. The link shown in Fig. 12 is one of the most commonly used in spatial mechanisms. The four points mass model illustrated in this figure has the following dimensions

$$
\begin{cases}
m_1 = m_2 = m_3 = m_4 = \dfrac{m}{4} \\[3mm]
x_1 = |x_2| = y_3 = |y_4| = i_{xx}\sqrt{2} \\[3mm]
|z_1| = |z_2| = z_3 = z_4 = i_{zz}
\end{cases}
\qquad (16)
$$

If the mass of a link is uniformly distributed along an axis, it could be concentrated in three points laying on the same axis. One of the points coincides with the point of gravity of the link and contains the 2/3 of its mass, while the other two points are located equidistantly on both sides of the point of gravity and these hold each a 1/6 of the link mass.

The model just described, simulates accurately the dynamical reactions of the commonly used links in manipulators (Fig. 1) and robotic hand/arm mechanisms.

When a solid performing a tridimensional motion is represented by a point mass system of a p number of concentrated masses, in order to establish the dynamical behaviour of this solid, i.e. to find out its inertia forces, it is necessary to know only the accelerations of the points of the system. The point masses and the respective accelerations define a discrete system comprising inertia forces, which need not to be reduced, for practical purposes, to a dynama of the inertia forces (resulting inertia force and resulting moment in a given point).

It should be noted that the determination of the inertia forces of a point mass model requires an investigation of the paths and the movement of single points of the solid, but not of the angle velocities and angular accelerations, this work being associated with additional calculation difficulties.

EXAMPLE

Today, hot forging probably represents one of the most hazardous environments in which production takes place under manual guidance. The dangers to the operators whilst handling hot materials and operating a hammer or a press are severe. Flying scale, high ambient temperatures, an atmosphere laden with dirt, lubricant spray etc. and the level of noise supply additional hazards to which the forging personnel are subjected [16].

We were faced with the problem to design a handling arrangement for the stamping operations of a technological cycle (Fig. 13) for the production of pinions for differential gearings (Fig. 14c), comprising: inductive heating 1 of billets, conveying the billets as heated to a stamping press 3, stamping the billets on the press in four operations, chuting the forged components to a trimming press 4, machining and chuting the end product. In Fig. 13, numerals 2 and 5, 6 designate a conveyer and two chutes, respectively. As shown in Fig. 15, the billets are cylindrical (In) and the three basic operations (I, II, and III), corresponding to the intermediate forms of the gear blank illustrated in Figs 14 a), b), c), respectively, are progressive die swaging operations, while operation IV is a sizing die one.

A first problem emerged in conjuction with the manipulation of the object in four positions (I, II, III, IV) with a view to fulfil all stamping operations simultaneously, as dictated by the press machine. Thus, we have proposed the system shown in Fig. 16 a) comprising five arms 1 - 5 provided with tailor-made grippers, the arms being rigidly fixed on a common yoke 6. Only the grippers of arms 4 and 5 are capable of rotary motions (secondary movements). The yoke 6 is controllable for three orthogonal translation movements: a first sagittal traverse with relation to the columns 7, a second vertical traverse defined by the movement of slider 8, and a third transversal traverse determined by the block 9 with relation to the base 10, secured on the stamping press. The provision of a fifth arm ensures the transfer of billet from the conveyer to position I, as well as the removal of the forged component (Ou), see Fig. 15.

Another problem was the centering of the gear blank in position IV: the teeth of the die and these of the blank should interpretate. A rotary movement of the gear blank between positions III and IV was sufficient to allow interpretation of the teeth by gravity. However, it was necessary to return the forged component in its initial attitude when removed from IV. Thus, a second rotary movement was the solution. As stated, both rotary movements are performed by the grippers of arms 4 and 5 (Fig. 16a)). In Fig. 16 b) we have shown the path movements of the handling arrangement.

REFERENCES

[1]   DRISCOLL L.C., Blue Collar Robots — A Technology Forecast, Proceedings of the Second International Symposium on Industrial Robots, IIT Research Institute, Chicago, Illinois, (May 16-18, 1972).

[2]   SHELDON O.L., Robots and Remote Handling Methods for Radioactive Materials, Proceedings of the Second International Symposium on Industrial Robots, IIT Research Institute, Chicago, Illinois, (May 16-18, 1972).

[3]   KLEINWÄCHTER H., The Antropomorphous Machine Syntelmann in Atomic Energy and Space Research, Proceedings of the Second International Symposium on Industrial Robots, IIT Research Institute, Chicago, Illinois, (May 16-18, 1972).

[4]   IGNATIEV M.B., KULAKOV F.M., POKROVSKII A.M., On the Problem of Deep-sea Manipulators with Automatic Control (in Russian), "Oceanology", 1970, No. 10.

[5]   MANCINI C., RONCAGLIA F., Il Servomanipolatore Electronica Mascot 1 del C.N.E.N. (in Italian), Alta Frequenza, 1963, No. 6, 32

[6]   GOTO T., TAKEYASU K., INOYAMA T., SHIMOMURA R., Compact Packaging by Robot with Tactile Sensors, Proceedings of the Second International Symposium on Industrial Robots, IIT Research Institute, Chicago, Illinois, (May 16-18, 1972).

[7]   VINOGRADOV I.B., KOBRINSKII A.E., STEPHANENKO Y.A., TIVES L.I., Salient Features in the Kinematic of Manipulators and Method of Volumes (in Russian), Mechanika Mashin, 1971, Issue 27-28.

[8] LINDBOM T.H., Today's Robots at Work in Industry: Matching the Robot and the Job, Proceedings of the Second International Symposium on Industrial Robots, IIT Research Institute, Chicago, Illinois, (May 16-18, 1972).

[9] HASEGAWA Y., An Approach to Industrial Robot Application Research, Proceedings of the Second International Symposium on Industrial Robots, IIT Research Institute, Chicago, Illinois, (May 16-18, 1972).

[10] "Industrial Robots – A survey" Booklet, International Fluidics Services Ltd., Bedford, England, 1972.

[11] JOHNSON K.G. , HANIFY D.W., The Current Status and Impact of Industrial Robot Technology in the United States, First Conference on Industrial Robot Technology, Paper R2, University of Nottingham, UK, March 27-29, 1973.

[12] KONSTANTINOV M.S., GENOVA P.J., BINEV A.M., A Strain Wave Transmitter with a Geometrically Optimised Flexspline, Proceedings of the Third World Congress for the Theory of Machines and Mechanisms, Kupari, Yugoslavia, September 13-20, 1971.

[13] HOCKEY B.A., The Method of Dynamically Similar Systems Applied to the Distribution of Mass in Spatial Mechanisms, Jnl. Mechanisms, Volume 5, pp. 169-180, Pergamon Press 1970.

[14] KOSTANTINOV M.S., GENOVA P.J., Dynamical Point Mass Models of Spatial Mechanisms, 12th ASME Mechanisms Conference, Paper No. 72-MECH-57, San Francisco, California, October 9-11, 1972.

[15] KOSTANTINOV M.S., GENOVA P.J., TOPENTCHAROV V., Système de Masses Concentrées Equivalentes à un Corps Solide, Annuaire de l'institut de mécanique appliquée et d'électrotechnique, Sofia, 1959, Vol. 6, No. 4, pp. 260-266.

[16] ROOKS B.W. , OKPERE K.O., CHENG R.M.H., Automatic Handling in Hot Forging Research, First Conference on Industrial Robot Technology, Paper R8, University of Nottingham, UK, March 27-29, 1973.

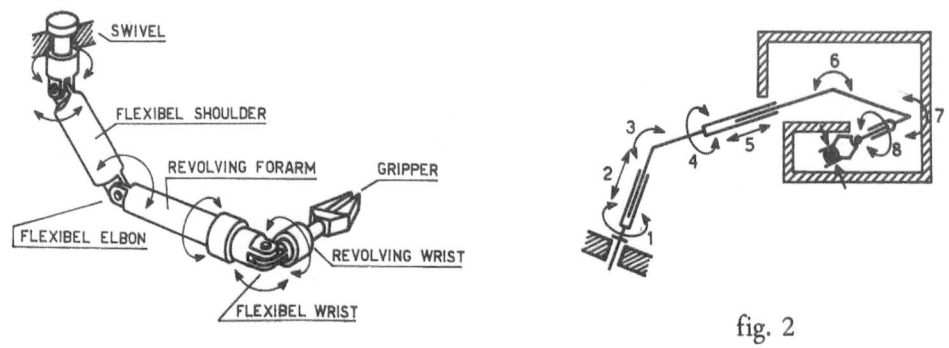

fig. 1

fig. 2

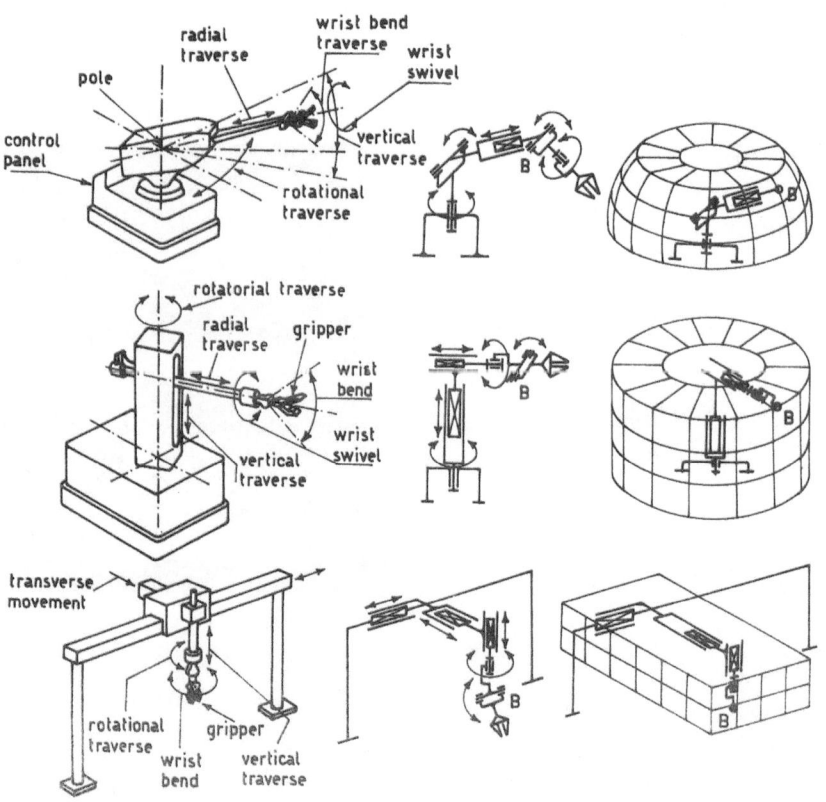

fig. 3

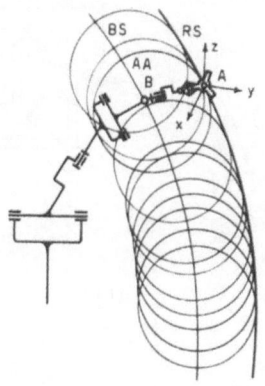

fig. 4

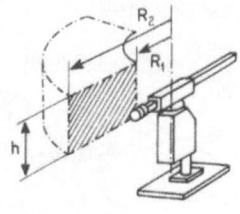

fig. 5

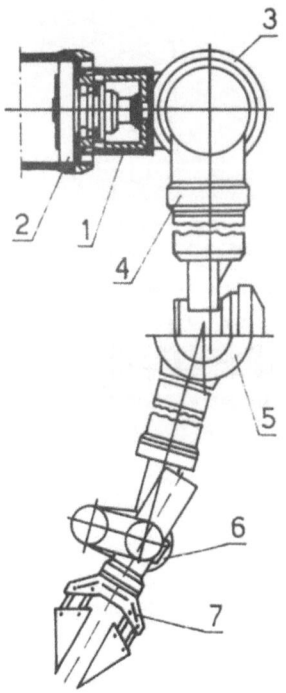

fig. 6

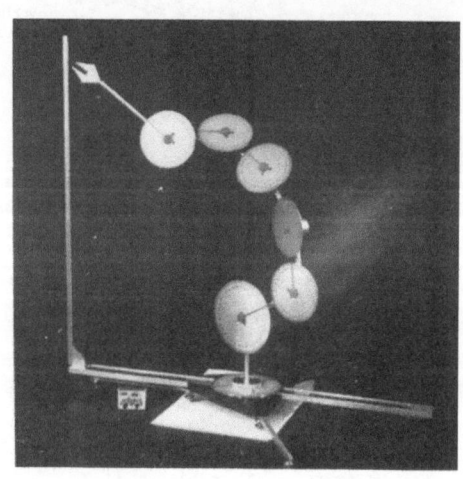

fig. 7

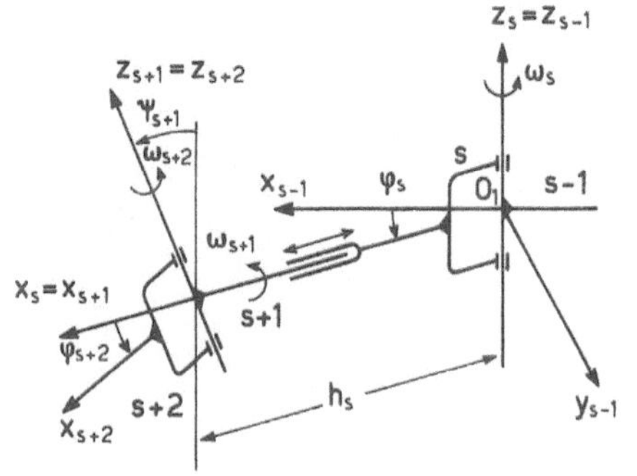

fig. 8

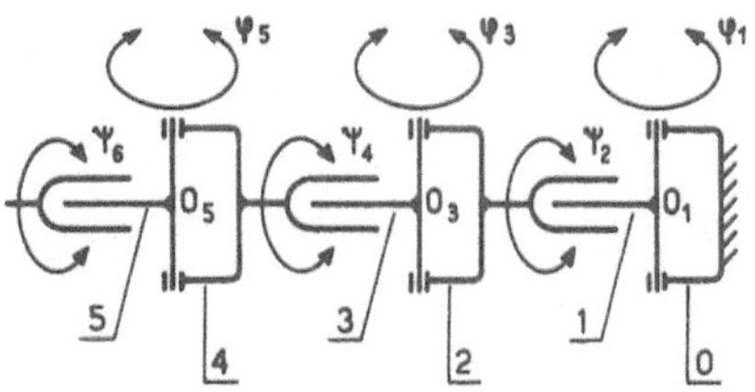

fig. 9

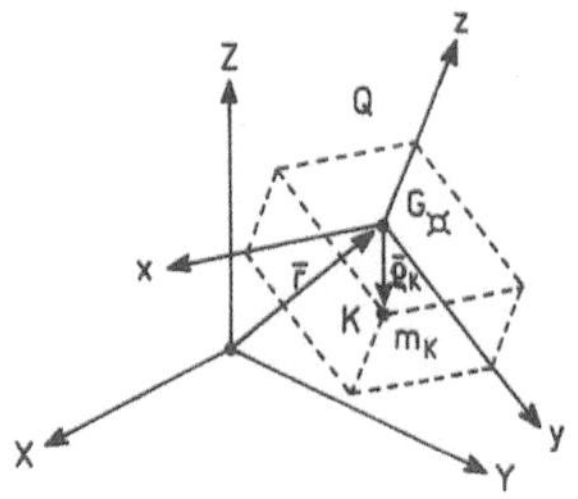

fig. 10

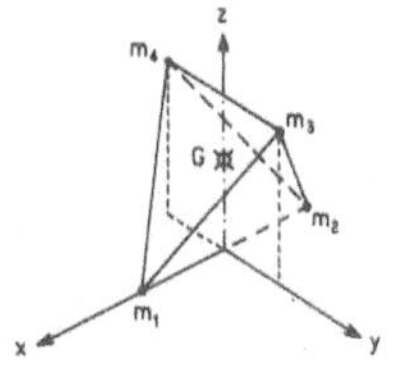

fig. 11

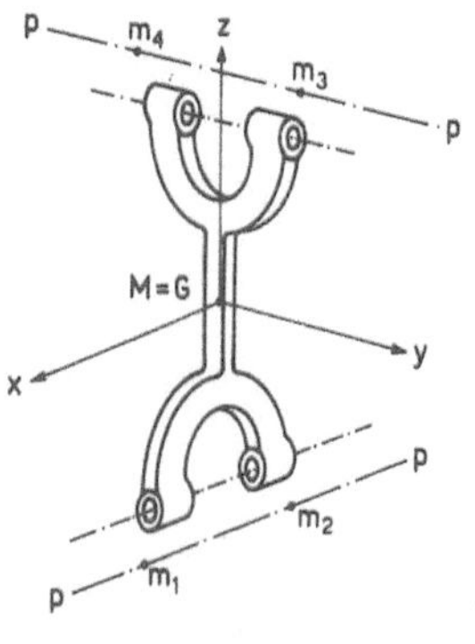

fig. 12

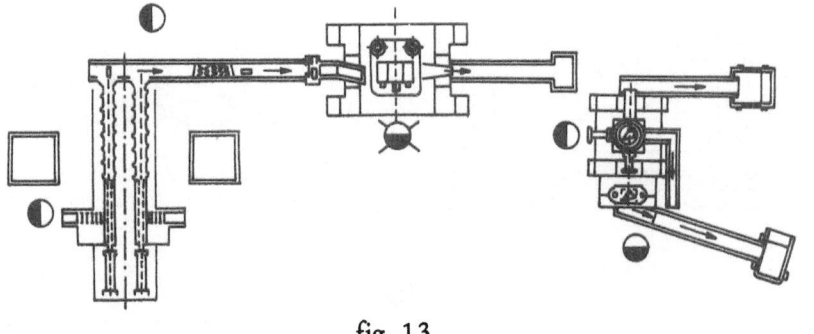

fig. 13

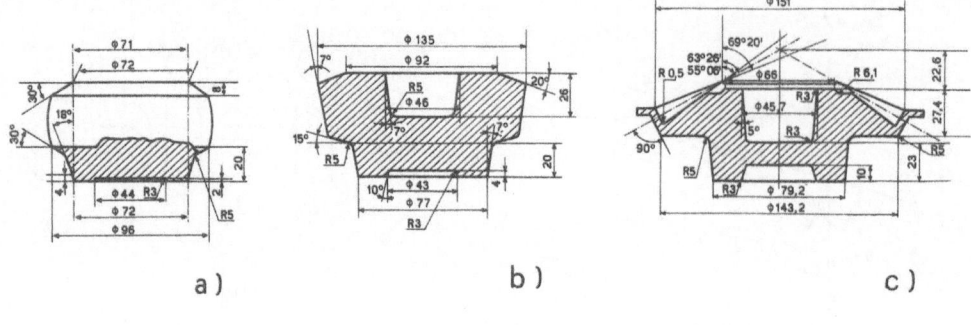

a)   b)   c)

fig. 14

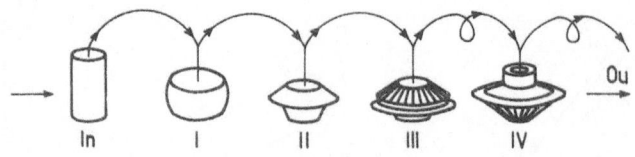

fig. 15

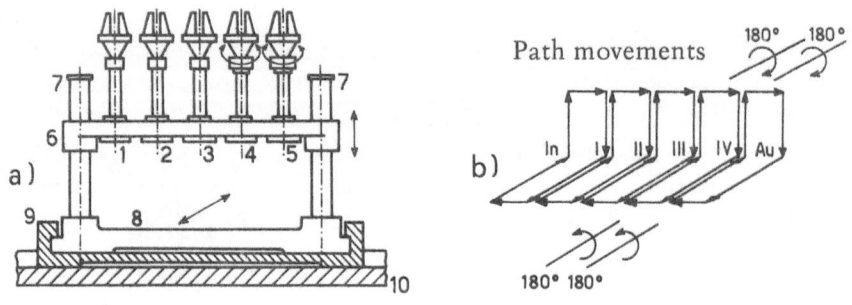

fig. 16

APPENDIX T A B L E 1 – SPECIFICATION OF INDUSTRIAL ROBOTS

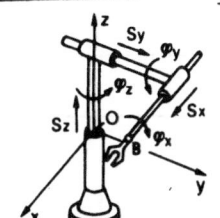

Primary movements

Sx – sagittal stroke
Sy – transverse stroke
Sz – vertical stroke
φx – rotation
φy – sweep
φz – swing

O – origin
B – wrist point

PTP – point-to-point
CP – continuous path
E – electric motor
H – hydraulic
P – pneumatic

Robot Name / Model	Country	Sx (mm)	Sy (mm)	Sz (mm)	φx	φy	φz
HAWKER SIDDELEY LTD. Minitran 700/90°	GB	–	–	50	–	–	180°
RHEIN-NADEL XH 30	GFR	200	300	300	300°	300°	300°
AIDA ENG AH-3	Japan	360	–	100	–	–	120°
B M Production Terminal	USA	350	250	250	–	–	90°
CORONA CPR-1	Japan	500	–	500	–	–	240°
ELECTROLUX MHU-Junior	Sweden	500	–	150	–	–	270°
R.KAUFELDT Robot A3	Sweden	750	–	300	–	–	270°
KOENIG Roboka 151	France	410	–	160	–	45°	–
ROBOTICS MIR-A-H-3	USA	1000	70	350	–	–	–
ANDROIDEN-GES. Man.System PTM	GFR	500	–	500	–	90°	90°
SUNDSTRAND Industrial Robot	USA	–	–	–	–	70°	330°
UNIMATION Unimate 2000	USA	1065	–	–	–	57°	220°
TRALLFA Trallfa-Robot	Norway	1000	–	2250	–	–	75°
TESA Tesamat 1	Norway	1000	–	800	–	–	360°
AMF VERSATRAN E 302	USA	750	–	750	–	–	240°
STAR SEIKI MHY-Type	Japan	1400	200	1100	–	–	–
HAYES DANA LTD. Robotex	Canada	1000	–	1000	–	–	180°
B + R TAYLOR Transiva	GB	609	–	305	–	–	240°
TAIYO Janbot-800	Japan	1150	–	325	–	–	180°
BURCH CONTROLS BCR 6000	USA	1200	–	1200	–	–	300°

α	β	γ
–	–	–
300°	–	–
180°	–	–
270°	–	–
–	–	–
provided		
–	–	–
–	195°	–
90°	–	–
–	–	–
360°	270°	–
300°	220°	200°
210°	210°	
360°	180°	–
180°	180°	–
–	–	–
280°	–	–
180°	–	–
–	–	–
–	–	–

a \ ^{pq}	0–1	1–2	0 – 2	1 – 3	0 – 3	1 – 4	
a_{11}	c_1	1	c_1	c_3	$c_1 c_3 - s_1 c_2 s_3$	c_3	
a_{12}	$-s_1$	0	$-s_1 c_2$	$-s_3$	$-c_1 s_3 - s_1 c_2 c_3$	$-s_3 c_4$	$s_1\,$
a_{13}	0	0	$s_1 s_2$	0	$s_1 s_2$	$s_3 s_4$	$s_1\,$
a_{21}	s_1	0	s_1	$c_2 s_3$	$s_1 c_3 + c_1 c_2 s_3$	$c_2 s_3$	
a_{22}	c_1	c_2	$c_1 c_2$	$c_2 c_3$	$-s_1 s_3 + c_1 c_2 c_3$	$-s_2 s_4 + c_2 c_3 c_4$	$-c_1\,$
a_{23}	0	$-s_2$	$-c_1 s_2$	$-s_2$	$-c_1 s_2$	$-s_2 c_4 - c_2 c_3 s_4$	$-c_1\,$
a_{31}	0	0	0	$s_2 s_3$	$s_2 s_3$	$s_2 s_3$	
a_{32}	0	s_2	s_2	$s_2 c_3$	$s_2 c_3$	$c_2 s_4 + s_2 c_3 c_4$	
a_{33}	1	c_2	c_2	c_2	c_2	$c_2 c_4 - s_2 c_3 s_4$	

a \ ^{pq}	1 – 5	0 – 5
a_{11}	$c_3 c_5 - s_3 c_4 s_5$	$c_5 (c_1 c_3 - s_1 c_2 s_3) + s_5 / s_1 s_2 s_4\,^{\cdot}$
a_{12}	$-c_3 s_5 - s_3 c_4 c_5$	$-s_5 (c_1 c_3 - s_1 c_2 s_3) + c_5 / s_1 s_2 s_4\,^{\cdot}$
a_{13}	$s_3 s_4$	$s_1 s_2 c_4 + s_4 (c_1 s_3 + \,$
a_{21}	$c_2 s_3 c_5 - s_5 (s_2 s_4 - c_2 c_3 c_4)$	$c_5 (s_1 c_3 + c_1 c_2 s_3) - s_5 / c_1 s_2 s_4\,^{\cdot}$
a_{22}	$-c_2 s_3 s_5 - c_5 (s_2 s_4 - c_2 c_3 c_4)$	$-s_5 (s_1 c_3 + c_1 c_2 s_3) - c_5 / c_1 s_2 s_4\,$
a_{23}	$-s_2 c_4 - c_2 c_3 s_4$	$-c_1 s_2 c_4 + s_4 (s_1 s_3 - \,$
a_{31}	$s_2 s_3 c_5 + s_5 (c_2 s_4 + s_2 c_3 c_4)$	$s_2 s_3 c_5 + s_5 (c_2 s_4 + \,$
a_{32}	$-s_2 s_3 s_5 + c_5 (c_2 s_4 + s_2 c_3 c_4)$	$-s_2 s_3 s_5 + c_5 (c_2 s_4 + \,$
a_{33}	$c_2 c_4 - s_2 c_3 s_4$	$c_2 c_4 - s_2 c_3 s_4$

a \ ^{pq}	1 – 6
a_{11}	$c_3 c_5 - s_3 c_4 s_5$
a_{12}	$s_3 s_4 s_6 - c_6 (c_3 s_5 + s_3 c_4 c_5)$
a_{13}	$s_3 s_4 c_6 + s_6 (c_3 s_5 + s_3 c_4 c_5)$
a_{21}	$c_2 s_3 c_5 - s_5 (s_2 s_4 - c_2 c_3 c_4)$
a_{22}	$-s_6 (s_2 c_4 + c_2 c_3 s_4) - c_6 / c_2 s_3 s_5 + c_5 (s_2 s_4 - c_2 c_3 c_4) /$
a_{23}	$-c_6 (s_2 c_4 + c_2 c_3 s_4) + s_6 / c_2 s_3 s_5 + c_5 (s_2 s_4 - c_2 c_3 c_4)$
a_{31}	$s_2 s_3 c_5 + s_5 (c_2 s_4 + s_2 c_3 c_4)$
a_{32}	$s_6 (c_2 c_4 - s_2 c_3 s_4) - c_6 / s_2 s_3 s_5 - c_5 (c_2 s_4 + s_2 c_3 c_4) /$
a_{33}	$c_6 (c_2 c_4 - s_2 c_3 s_4) + s_6 / s_2 s_3 s_5 - c_5 (c_2 s_4 + s_2 c_3 c_4) /$

pq a	0 – 6
a_{11}	$c_1 (c_3 c_5 - s_3 c_4 s_5) - s_1 / c_2 s_3 c_5 - s_5 (s_2 s_4 - c_2 c_3 c_4)$
a_{12}	$c_1 / s_3 s_4 s_6 - c_6 (c_3 s_5 + s_3 c_4 c_5) / + s_1 c_6 / c_2 s_3 s_5 + c_5 (s_2 s_4 - c_2 c_3 c_4)$
a_{13}	$c_1 / s_3 s_4 c_6 + s_6 (c_3 s_5 + s_3 c_4 c_5) / - s_1 s_6 / c_2 s_3 s_5 + c_5 (s_2 s_4 - c_2 c_3 c_4)$
a_{21}	$s_1 (c_3 c_5 - s_3 c_4 s_5) + c_1 / c_2 s_3 c_5 - s_5 (s_2 s_4 - c_2 c_3 c_4)$
a_{22}	$s_1 / s_3 s_4 s_6 - c_6 (c_3 s_5 + s_3 c_4 c_5) / - c_1 c_6 / c_2 s_3 s_5 + c_5 (s_2 s_4 - c_2 c_3 c_4)$
a_{23}	$s_1 / s_3 s_4 c_6 + s_6 (c_3 s_5 + s_3 c_4 c_5) / + c_1 s_6 / c_2 s_3 s_5 + c_5 (s_2 s_4 - c_2 c_3 c_4)$
a_{31}	$s_2 s_3 c_5 + s_5 (c_2 s_4 + s_2 c_3 c_4)$
a_{32}	$s_6 (c_2 c_4 - s_2 c_3 s_4) - c_6 / s_2 s_3 s_5 - c_5 (c_2 s_4 + s_2 c_3 c_4)$
a_{33}	$c_6 (c_2 c_4 - s_2 c_3 s_4) + s_6 / s_2 s_3 s_5 - c_5 (c_2 s_4 + s_2 c_3 c_4)$

Digital schemes

0-1	= 2-3	= 4-5		
	0-2	= 2-4	= 4-6	
		0-3	= 2-5	
			0-4	= 2-6
				0-5
				0-6

1-2	= 3-4	= 5-6	
	1-3	= 3-5	
		1-4	= 3-6
		1-5	
			1-6

T A B L E 3

Mascot 1	
$\varphi_1 = 0 \quad s_1 = 0 \quad c_1 = 1$ $\Psi_4 = 0 \quad s_4 = 0 \quad c_4 = 1$	

$\bar{I} = \bar{I}_3 + \bar{I}_5 = \bar{\lambda}_3 h_3 + \bar{\lambda}_5 h_5$ $\quad \bar{\omega}_{60} = \bar{\lambda}_1 \dot{\Psi}_2 + \bar{\nu}_3 (\dot{\varphi}_3 + \dot{\varphi}_5) + \bar{\lambda}_5$

$\bar{I} = \bar{\lambda}_o (h_3 c_3 + h_5 c_{3+5}) + \bar{\mu}_o c_2 (h_3 s_3 + h_5 s_{3+5}) + \bar{\nu}_o s_2 (h_3 s_3 + h_5$

$\dot{\bar{I}} = \bar{V} = - \bar{\lambda}_o / h_3 s_3 \omega_3 + h_5 (\omega_3 + \omega_5) s_{3+5}/$

$\qquad + \bar{\mu}_o / h_3 c_2 c_3 \omega_3 + h_5 (\omega_3 + \omega_5) c_2 c_{3+5} - s_2 \omega_2 (h_3 s_3 + h_5$

$\qquad + \bar{\nu}_o / h_3 s_2 s_3 \omega_3 + h_5 (\omega_3 + \omega_5) s_2 c_{3+5} + c_2 \omega_2 (h_3 s_3 + h_5$

$\bar{\omega} = \bar{\lambda}_o (\omega_2 + \omega_6 c_{3+5}) + \bar{\mu}_o / \omega_6 c_2 s_{3+5} - (\omega_3 + \omega_5) s_2 / +$

ON THE DESIGN OF COMPUTER CONTROLLED MANIPULATORS

Bernard ROTH

Professor, Department of Mechanical Engineering

Jahangir RASTEGAR

Graduate Student, Department of Mechanical Engineering

Victor SCHEINMAN

Research Associate, Department of Computer Sciences

Stanford University
Stanford, California, U.S.A.

(*)

Summary

      In connection with researches in the area of Artificial Intelligence, several studies have been made of mechanical manipulators for use under computer control. This paper summarizes these results, and gives the reader an overview of problems associated with the design, kinematics and control of mechanical manipulators. Some new results on kinematics are presented.

(*) All figures quoted in the text are at the end of the lecture

Introduction

A man uses his hands to manipulate his environment in order to accomplish a multitude of different tasks. Some of these tasks rely exclusively on tactile or sensory feedback, and others are primarily motion manipulations which simply require a spatial displacement. Almost all manipulations combine the tactile and displacement modes, but in many cases one predominates, the other being secondary or even negligible. In this paper we will be mainly concerned with displacements, since it is in this realm that man tends to interact with machines and inanimate objects, and it is in this realm that we wish to replicate his abilities. This paper summarizes work done by the authors and their associates, and attempts to give an overview of the problems associated with the design, kinematics, and control of mechanical manipulators for use under computer control in connection with studies in Artificial Intelligence.

In undertaking the design of a machine with the ability to accomplish tasks similar to the ones performed by humans, one is immediately confronted with the choice of whether of not to attempt to replicate the human. Furthermore, the system aspects of the problem become immediately clear. In manipulations the hand cannot be taken as an isolated entity for it must rely on at least the arm and shoulders to assist it in most displacements. Similarly, the brain and nervous system enter into the "design" of the human hand. This latter idea is best illustrated by the widely held belief that man's evolutionary development of an opposable thumb was strongly interconnected with the development of his brain.

Based on these and similar thoughts we conclude that there is in fact no a priori reason to attempt to construct a mechanical manipulator which replicates the human hand-arm system (even if one could, and we cannot), until the brain and nervous system can also be replicated. Instead, in considering the design of a manipulator to be used in tasks usually done by humans, the machine must be considered as a whole. In particular it should be known if it is to be used as part of a man-machine system (such as a prosthetic device, or master-slave manipulator) or is it to be used in a completely automatic system (such as a robot, or industrial manipulator).

In this paper we restrict ourselves to the completely automatic group. In order to proceed with a manipulator design we must know about the rest of the system: Is there a computer which will direct the manipulator ? Is it digital, analog, or hybrid ? Is it a general or special purpose computer ? All these factors are relevant and so bias the manipulator design. Equally important is the manner in which the computer is to direct the manipulator. It is quite a different matter if the computer is to store a table of manipulator operations or generate its own sequence of motions.

Another area of interest are the sensors. Is the system able to get information from its environment ? Can it detect light or sound waves ? Is it sensitive to odors ? Can it sample the mass or rigidity of objects ? All of these factors may have a marked influence on the design. The situation is analogous to people tending to develop in ways which compensate for certain physical anomalies or sensory defects.

All manipulators require some form of controller. Are the controllers to be built-in ? Are reflex type responses required ? Is the computer to act as the controller, or in conjunction with it ? What type of control is desirable ?

Probably the most important factor in the design is the use to which the manipulator is to be put: Does it have to perform a rather limited number of tasks ? Will it have to act slowly or quickly ? Will it require great precision ? Will it have to make large displacements ? Will it have to exhibit great dexterity ? How large are the objects to be manipulated ?

In each of the above categories there are many more questions which need to be asked . But by now, hopefully, the reader has the thrust of the argument. Similarly, there are other categories which might be included such as life, cost and so on, but these are common to all designs. Assuming we have the answers to all these questions we can now consider the design of the manipulator. We begin by discussing the possible kinematic configurations.

Kinematics

A manipulator is generally built in the form of an open loop kinematic linkage(*). The number of kinematically possible configurations is unlimited. It is convenient to think of the linkage as having a fixed and free end. The freedom of the chain is equal to the degrees-of-freedom of the free end relative to the fixed one. Theoretically the freedom should be at least six, since this is the minimum required to place the free end at an arbitrary point with an arbitrary orientation within the

---

(*) It is conceivable to design manipulators which have closed loop subchains. These are however not used practically. Formally, such manipulators could easily be included in our development.

work space. In practice, however, more limited manipulators with only five
d.o.f.(**) are not uncommon. A recent survey of 17 manipulators [1] (***) revealed
six with five d.o.f., four with six d.o.f., two with four d.o.f., one each with three,
seven, and eight d.o.f., and two with more than eight d.o.f. In addition, most of
these were fitted with terminal devices which allow for holding objects.

Although the d.o.f. give one form of kinematic classification it is not
sufficient; for manipulators with the same d.o.f. may be quite different kinematically. It has been shown [1,2] that an adequate classification scheme can be
developed based on the type and order of the joints (which incidentally yields the
d.o.f.), and a sub-classification according to which adjacent joint axes do not
intersect and which adjacent common normals to the joint axes do not intersect.
According to this classification it is possible to divide all manipulators into solvable
and unsolvable types. This distinction is somewhat artificial but it may be important
in certain applications.

In the kinematic analysis of manipulators two problems may arise. The
simple computation is the one where all the relative joint displacements are known
and the position of every link including the free end are to be computed. This
problem can always be solved by a straight-forward application of the well-known
Matrix method of kinematic analysis. The more difficult position problem is the one
where the question of solvability enters. Here the position and orientation of the
free end are specified, and it is required to determine a set of relative joint
displacement which correspond to the given position of the end. If an algorithm
exists for always determining all possible joint displacements which can bring the
manipulator free end into any required position within the work space, the
manipulator is said to be solvable. It has been shown [1,2] that, for manipulators
with revolute and prismatic joints, a sufficient condition for solvability is that any
three adjacent joint axes intersect in a common point. Solvability is useful in
systems where the computer must, by working from the desired coordinates of the
free end, direct the manipulator from one configuration to another.

In systems where a given manipulator is led through a series of motions
which are then stored for later use, solvability is not important. Similarly, systems

---

(**) We abbreviate degrees-of-freedom with d.o.f.

(***) Numbers in square brackets refer to references listed at the end of the paper.

which work by moving with small incremental motions in a preferred direction do not require solvability. However, in these systems it is useful to have the various d.o.f. decoupled enough to allow a displacement in any direction from any given position. In practice it turns out that most such systems in which the freedoms decouple are also solvable.

Although the question of solvability was considered in terms of manipulators with 6 d.o.f. the results given in [1,2] are applicable to manipulators with any number of freedoms. Manipulators with more than 6 d.o.f. may be treated as 6 d.o.f. systems with the extra freedoms taken as arbitrarily specified or as free parameters. Manipulators with fewer than 6 d.o.f. are automatically included as special cases of the 6 d.o.f. systems.

When manipulators are not solvable or small incremental motions are desired two schemes have been found useful. In both, large displacements are obtained by breaking them up into a series of small ones. The less efficient method uses the well-known Newton-Raphson iteration, a better algorithm follows from an iteration based on the use of velocity screws [1].

In order to develop the Newton-Raphson method, the displacement equations must first be written. These are always obtainable in a straightforward manner by use of the well-known Matrix methods of kinematic analysis. The basic assumption in the Newton-Raphson technique is : If γ_{io} represents the relative displacement in the ith joint for a known manipulator position, any new adjacent configuration may be obtained by an additional relative joint displacement $\delta\gamma_i$ which can be obtained by solving the linear set of Matrix equations

$$[A] \begin{bmatrix} \delta\gamma_1 \\ \delta\gamma_2 \\ \vdots \\ \delta\gamma_6 \end{bmatrix} = [B] \tag{1}$$

for the $\delta\gamma_i$. Here [A] and [B] are 6x6 matrices which depend on the manipulator structure geometry, γ_{io}, and the required position and orientation of the free end of a 6 d.o.f. manipulator. When [B] = 0 the manipulator is in the desired position. Equation (1) is obtained by substituting $\gamma_{io} + \delta\gamma_i$ for unknown joint displacements in the displacement equations and neglecting all non-linear terms in the $\delta\gamma_i$ which are assumed to be infinitesimal quantities. Each time equation (1) is used to compute the $\delta\gamma_i$ their values are checked to see if they are small enough to consider the present position of the free end as the desired one. If they are not small enough the computation is repeated using a new number for γ_{io}. The new γ_{io} is the previous value of the γ_{io} plus the present value of $\delta\gamma_i$. The procedure is repeated until the corrections $\delta\gamma_i$ get small.

The velocity method requires writing the velocity of the free end in terms of the summation of the velocities due to the several joint displacements. Then taking the displacements as being small and replacing the time derivatives by ratios of infinitesimals, the incremental change in time for the displacement can be eliminated. The result is two sets of linear vector equations from which the six scalar joint displacements $\delta\gamma_i$ can be computed:

$$\sum_{i=1}^{6}[(\delta\gamma_i)\underline{n}_i] = (\delta\omega)\underline{n}$$

(2)

$$\sum_{i=1}^{6}[(\delta\gamma_i)(\underline{n}_i \times \underline{r}_i)] = \underline{n} \times \underline{r} - p\delta\omega\underline{n}$$

Here we assume a 6 d.o.f. revolute joint manipulator.

$\underline{n}_i$ is the unit direction vector of the ith joint before the displacement.

$\underline{r}_i$ is any position vector from an arbitrary origin to the ith joint axis before the displacement.

$\underline{n}$, $\underline{r}$, p, $\delta\omega$ are respectively the unit direction vector, the position vector from the origin, the pitch, and the rotation angle associated with the screw axis for the displacement which brings the free end from its present position to the desired new position.

The iteration that follows from equation (2) uses a present value of $\underline{n}_i$ and $\underline{r}_i$ to determine a new set of $\delta\gamma_i$. Once the $\delta\gamma_i$ are known, the manipulator is considered as having a new present position, and new $\underline{n}_i$ and $\underline{r}_i$ may be computed, as well as new values for $\delta\omega$, $\underline{n}$ $\underline{r}$ and p. This procedure is repeated until the right sides of equations (2) are small enough for the particular manipulator application.

This velocity method seems to be generally superior to the Newton-Raphson method. For example, Pieper [1] give the following test results: for free end position and orientation changes which required joint angle changes of $30°$, $20°$, $-10°$, $30°$, $-10°$, $-30°$ in the six revolute joints (respectively from the fixed to the free end) of a 6 d.o.f. manipulator, the velocity method required 10 iterations in 0.97 seconds. The Newton-Raphson technique required 13 iterations in 1.82 seconds to arrive at a solution which was slightly less accurate. Using a PDP-6 computer and FORTRAN IV code it was found that on the average the velocity method ran twice as fast as the Newton-Raphson procedure. Furthermore, and most important, it converged in all cases tested, whereas the Newton-Raphson procedure did not. For small displacements the velocity scheme is extremely efficient with regard to computation time. Using this method a general purpose manipulator solving program has been developed. Given the position of a manipulator and a desired new position and orientation of the free end, the program computes the required joint displacements for any 6 d.o.f. manipulator with revolute or revolute and prismatic joints.

In general, the computation time required to determine joint displacements are much shorter for solvable arms. However, iterative techniques are almost always faster when the required distance and orientation changes are small and only one configuration in the final position is required. In order to avoid obstacles it is, however, often necessary to know more than one possible configuration corresponding to the same free-end position. Solvability therefore is a very desirable characteristic.

Even if manipulators are solvable certain positions will generally not be possible. This will result in the computation yielding imaginary numbers for the numerical values of joint displacement, or non-convergence in iterative solutions. Clearly there is a certain limit to the distance a manipulator can reach outward from the fixed end, and a limit to the amount it can fold back upon itself. Even if a position can be reached with the terminal end it may be impossible to obtain a desired orientation. It is, however, possible to guarantee a full range of orientation freedom by making the last three joints (at the free end) intersecting revolutes each capable of 360° rotation. The question of zones of action is considered briefly by Pieper who also gives a tabulation of orientation resctrictions for 120 different six-revolute manipulators [1].

Number of Solutions

A question of both theoretical and practical interest that does not seem to have been previously treated is:

For any given manipulator, how many configurations yield the same position and orientation of the terminal device ?

If we know the answer to this question we can decide when we have found all possible configurations, and determine which configuration is best in the sense of auxiliary constraints (e.g., avoiding obstacles, or obtaining minimum displacements in certain joints).

The most general 6 d.o.f. manipulator with all turning joints, is equivalent to the 7 revolute-joint (i.e., 7R) spatial seven-bar linkage. In order to study it, it is convenient to first consider the spatial six-link 5R,C linkage (i.e., one cylindric and 5 turning joints). A skeleton diagram of this linkage is shown (Schematic 1a).

Since the input-output equations for this linkage are of 16th degree [8,9], it follows that for any given position of link 5 there are at most 16 positions of point 0_5 (attached to link 4) which allow this linkage to be assembled. These 16 positions all lie on line S_5. Now, if we consider R_5 to be free, the position of line S_5 is arbitrary. Since 0_5 intersects arbitrary line S_5 in sixteen points we conclude that the locus of all possible positions of 0_5 is a 16th order surface. Now, if we replace the cylindric joint and link 5 with the links 5 and 6 shown in Schematic 1b we conclude that point 0_5^1 lies on a circle. Since a circle intersects a 16th order surface in at most 32 points we conclude that 0_5 and 0_5^1 can coincide in at most 32 positions. This means then that there are at most 32 configurations (for a 6R manipulator) corresponding to a position and orientation of the terminal device. Furthermore, this implies a new result in kinematic analysis: the degree of the input-output equations for a general 7R spatial linkage cannot even be less than 32nd.

If joints are replaced by sliders the number of possible configurations will be less than 32. For example consider the 5R,P manipulator (which is equivalent to a 5R,P,R seven-bar closed linkage). The argument is as in the case of turning joints, except that we replace the cylindric joints by the revolute and prismatic pair shown in Schematic 1c. Here the locus of 0_5^1 is a line parallel to S_6 (we consider link 6 fixed). Therefore 0_5 and 0_5^1 can only coincide in the 16 points of intersection of this line and the 16 th order surface. In this case we have at most 16 manipulator configurations for each position and orientation of the terminal device. Furthermore, the degree of the input-output equations of any 6 revolute, 1 prismatic joint spatial seven-bar linkage cannot be generally less than 16.

The equations implied by these synthetic arguments will be discussed elsewhere.

We close this section by pointing out that the distinction between solvable and not solvable manipulators is not an invariable one. In fact, if the 32 degree equation characterizing the 6R manipulator can be formulated and solved numerically, all 6 d.o.f. revolute-joint systems will become solvable.

Control and Dynamics

We now shift our attention from the kinematics to the dynamics and control of mechnical manipulators. The problems in this area are strongly dependent on the rates of motion. If the manipulator is to be moved very slowly no significant dynamic forces will act on the system. However, if rapid motions are desired dynamic forces become high, dynamic coupling becomes significant, dynamic unbalance and vibrations become problems.

The fundamental question in manipulator control can be stated as follows: what forces or torques must be applied at the manipulator joints in order to move the manipulator from its present configuration to the desired one ? In addition, ther may be ancillary conditions on the control such as minimizing the time or power for the motion, or keeping certain acceleration below a given level.

The simplest solutions to control problems exist when each joint can be controlled independently. This is the case with very slow motions . Here simple proportional control (where the force applied to each joint is a function of the required displacement of the joint) often suffices. If rapid motions are desired a more complex control system is required since the dynamic interactions between the links must be taken into account. Before considering the characteristics of possible control systems we examine some design criteria based on dynamic effects.

After kinematic criteria have been used to specify the geometry of a manipulator, dynamic characteristics must be considered. Most important are the inertia characteristics and torque or force bounds required to meet response time specifications. In general these specifications will lead to torque/inertia ratios for the joints of the kinematic chain. Due to dynamic coupling and varying inertia loads, the analysis of determining these ratios in an actual manipulator is generally rather complex. There does not seem to be a general non-iterative scheme. It remains then simply to assume values for unknown parameters and then study the response of the system. If the design criteria are not met some of the parameters may be changed and the process repeated. It may be useful to keep the following points in mind:

a) The inertia loads effectively acting at the joints can be minimized if the body masses and lengths decrease successively from the fixed end outward.

b) Torque and force bounds must be selected to satisfy torque/inertia or force/inertia ratios determined from the required response time.

c) The gravity loads and loads due to the angular velocities of the members should be small compared to the maximum available torques and forces.

d) Dynamic and static loads must be taken into account in deflection analysis if very fast response times are required.

The equations of motion for any 6 d.o.f. manipulator with revolute joints have been derived and listed in their most general form [3]. Using these equations it becomes possible to determine the dynamic loads as well as to study the question of controlling a rapidly moving manipulator.

A proportional control is the simplest closed loop control to implement. In its most general form it is constructed so that at any time t the torque or force T_i applied to the ith joint by the controller is given by

(3) $$T_i = - g_i + k_{pi} \beta_i + k_{vi} \dot{\beta}_i + k_{Ii} \int \beta_i \, dt$$

where $\dot{\beta}_i$ is difference between the position at t and the desired joint position
β_i is the difference between the velocity at t and the desired joint velocity
g_i is the gravity force or torque at t

k_{pi}, k_{vi}, k_{Ii} are gain constants for the position, velocity and integral components respective.

In practice the constants must be adjusted to obtain sufficiently fast by well damped response. The integral term in equation (3) is useful in eliminating position errors due to small long-term disturbances such as leakage in hydraulic actuators and other "drifts".

The above is an ideal proportional controller. In practice there will generally be noise and other errors in determining position and velocity. To compensate for this two adjustments are useful. The first involves setting β_i to zero whenever its magnitude falls below a certain absolute value. This allows for the use of reasonably large velocity damping, and for the removal of this source of noise in the control signal at steady-state. The second adjustment involves using a weighted combination of the current and most previous reading of position and velocity instead of β_i, and $\dot{\beta}_i$ respectively. This, too, helps to smooth out the effects of noise.

If it is desired to optimize the control in some way, additional constraints must be considered. In most cases the porportional control will not be the optimum one. It is, however, almost always the simplest. In the rest of this section we consider the problem of designing the controller so as to carry out all the motions in minimum time.

It is well known that bang-bang control is required for minimum-time motions. The problem of designing a minimum-time controller is then one of determining when to reverse each control during a motion. In general the dynamical equations of the system are nonlinear and a closed-form representation of the minimum-time feedback control is not possible. It does however seem feasible to develop suboptimal controls which provide close approximations to the optimal control.

It can be shown [3,4] that for manipulators whose equations of motion can be inverted into the form

$$\dot{\underline{\Phi}}(t) = \underline{f}[\underline{\Phi}(t), \underline{T}(t), t] \tag{4}$$

where: $\underline{\Phi}(t)$ is a vector with 2n components representing the state of the system at time t

$T(t)$ is a vector with n components denoting the control input (e.g., torque) to the system at time t

$\underline{f}$ [ ] is a vector-valued function of the state $\underline{\Phi}$, control T, and time t

the minimum principle of Pontryagin furnishes local necessary conditions which optimal control must satisfy. This then yields the condition that a Hamiltonian dependent on $\underline{f}$ be minimized at every instant during the motion. In order to carry out this minimization it is necessary to solve a rather difficult two point boundary value problem. It is at this point that an approximation to the optimal control must be introduced.

An approximate or so-called "suboptimal" control can be derived as follows: After a change of variables, the equations of motion are linearized and a transformation derived to uncouple the controls in the linearized equations. A switching surface accounting for the gravity loads on the system is then obtained and an approximation is made for the effects of the angular velocity terms in the nonlinear system. These approximations are used to obtain effective torque bounds and switching curves for each control. As will be described in the section on case studies this method was successfully tested on a 3 d.o.f. manipulator with hydraulically driven revolute joints 3,4 .

Use of a digital computer in the suboptimal control feedback loop requires some modification of the switching curves to account for the manipulator motion between data sampling instants. An additional modification to prevent steady-state oscillations near the origin of the solution space requires replacing the bang-bang with a proportional control near the very end of the motion.

Design

Given the foregoing considerations, is there an ideal manipulator configuration ? As in most design problems there does not seem to be any one correct answer. However, we are now in a position to critically compare various alternatives. We start by considering the joints.

In general, joints must be individually powered and controllable, and allow for simple relative motions between the members they connect. On these grounds revolute and prismatic joints are the most desirable. A cylindric joint may be considered as a special combination of a revolute and a prismatic joint, and does not need to be considered separately. Helical (or screw) joints are usually not useful because they give a fixed coupling between rotation and translation. Spherical joints are generally difficult to use. They have the disadvantage of allowing for rotations about axes whose directions vary with respect to the directions of the torques powering the joint. When the kinematic effect of a spherical joint is desirable it is usually best to use three mutually intersecting revolutes which are kinematically equivalent. Similarly a sphere-and-pin 2 d.o.f. joint is best constructed in the form of two intersecting revolutes. The above seem to be the joints which are generally used in manipulator design. In cases where the manipulator does not have a fixed end its contact with ground is usually a plane-on-plane 3 d.o.f. "joint". Here we approach an entirely new area beyond the scope of this paper, since we consider only manipulators with one end "fixed".

Manipulators with all revolute joints are in general the easiest to construct. They have the further advantage of being rather flexible in being able to reach around objects. Such manipulator can always be made in solvable configurations by having three adjacent mutually intersecting axes. They are widely used as artificial arms and in exoskelital bracing. Indeed, revolute manipulators do generally parallel the performance of human arms: They are good for throwing, but not for drawing extremely accurate straight lines or following arbitrary paths. The displacement errors at the free end increase with the manipulator in the stretched out position. Hence, precision work must generally be done close to the fixed end. Revolute manipulators tend to be difficult to counterbalance, and they have high rotary inertias. These effects may be reduced by transmitting power to the joints via belts, ropes, cables, or chains, instead of mounting the motors directly on the joints.

Manipulators with three prismatic and three revolute joints are solvable regardless of how the joints are combined. Such manipulators can be designed to give uniform precision over the entire operating range. They can be made rather flexible if revolute and prismatic joints are alternated. In practice, the three prismatic joints are usually placed adjacent to each other at the fixed end. Such manipulators are able to easily avoid obstacles by going around them, but are less able to reach into-and-around enclosed spaces. They lend themselves nicely to rigid construction, and, because of fairly uniform gravity loadings, have smaller peak power requirements than other types of manipulators. Such configurations are often found in overhead crane type manipulators. They are ideal for following a given path with their free end.

Using only one or two prismatic joints results in characteristics which combine the advantages of the two previous types. Such manipulators tend to have the lowest weight and minimum structure for a given useful working volume. They also tend to have a larger workspace for a given manipulator size, and shorter working times because of smaller joint displacements to obtain a given end position. In terms of obstacle avoidance, path following, and such tasks as throwing, these manipulators seem to be midway between the extremes of the previous two cases.

In general manipulators with more than 6 d.o.f. are desirable since they are more versatile, especially if some of the joints can be "locked" when not in use. However, such manipulators tend to be more expensive, since each additional joint requires a power source, controls, feedback sensors, and computer time.

For any configuration, the range of motion of a joint should be maximized. For example, a revolute joint which turns through 300°, as opposed to the entire 360°, may greatly reduce the utility of a manipulator. This is especially true in applications requiring flexibility such as obstacle avoidance.

In order to avoid mechanical interference and allow a manipulator to work in small openings, a slim outline is desirable. This also permits a manipulator to be easily recognized by an optical sensing system. If the manipulator position is controlled by sensing the free end by optical 6 or other means, the deformation of the structure may not be important. However, since most manipulators rely on position feedback from the joints (via potentiometers or shaft encoders), the structural members must not be slim enough to have their rigidity adversely affected. This then mitigates in terms of high stiffness to weight ratio sections, such as large diameter round and square tubes. It is also advantageous to use large diameter, thin section ball bearings.

The three common power sources are electrical, hydraulic, and pneumatic. Electricity is universally available and inexpensive. Since most applications use d.c. motors, rectified or battery sources are needed. Hydraulic power provides a large amount of energy with a minimum actuator weight, but requires an expensive power supply and control system. Pneumatically powered manipulators are cleaner and cheaper than hydraulic systems. However, for safety reasons, they must operate at much lower pressures and therefore have poorer dynamic response.

Position sensing should use highly linear elements with high resolution. Nonlinearity may be tolerated if computer memory is available to store calibration curves and time is available for table lookup routines. In general gear trains between the joints and the position measuring device should be avoided in order to minimize blacklash and nonlinearities.

Case Studies

An extensive design study of computer controlled manipulators has recently been carried out by one of the authors [5]. The resulting manipulator, shown in Figure 1, has one prismatic and five revolute joints. The prismatic joint is the third from the fixed end. The terminal device has a seventh freedom in the form of a parallel jaw vice grip. This manipulator has been constructed and is presently being successfully used in connection with the Stanford Artificial Intelligence Project.

This manipulator is powered by permanent magnet d.c. motors with "Harmonic Drive" speed reducers. Bi-directional electromagnetic breaks are used on the shafts. The motors are driven under direct digital control as slaved pulse following servo motors. Large length potentiometer elements with integral wipers are directly mounted on the joints. Twelve bits resolution (one part in 4096) are provided for the full scale 360° rotation of the two largest revolute joints . The manipulator has provision for a six component strain gage for force measurement at the free end.

A new study recently completed by Scheinman has resulted in plans for a more compact design. Earlier two rather different manipulators have been designed and constructed at Stanford University. One is a hydraulically powered, 6.d.o.f., manipulator with revolute joints, and one additional freedom in the form of an air driven terminal jaw (Fig.2). The second is a pneumatically driven manipulator with revolute joints designed to be limited to a discrete set of configurations (Fig.3).

The hydraulic arm, Fig. 2, was used to test the suboptimal (near-minimal time) controller. In these tests the three joints nearest the free end were held locked in their zero position, so the controller was working with, in effect, a 3 d.o.f. system. Torque was applied to each joint through specially designed rotary hydraulic actuators. The actuators were controlled by two-stage position feedback servovalves. Feedback from differential pressure transducers on the actuators allowed for input current to the valves to control actuators output torque. The maximum torque on the first two joints was 2000 psi, while the third had 1000 psi.

Since the tests were run under time-sharded conditions, it was necessary to modify the suboptimal control for use in a sampled-data system. It was found possible, by shifting the switching curves, to obtain satisfactory results even at slow sampling rates. Test results were found to agree well with simulations of the ideal system [3]. The movie, accompanying the oral presentation of this paper, shows the system being tested. The sampling rate is 60 Hz.

Figure 4, taken from [3] shows the results of one such test. ζ_i represents the angular rotation of the ith joint; the required final position in each joint is $\zeta_i = 0$. In addition to displacement versus time curves for the three joints, actual actuator torque is compared to that predicted by an idealized simulation of the system. Interestingly, in many cases, the dynamic reaction forces were large enough to require the control to act opposite to the direction in which the motion was required.

The manipulator shown in Fig. 3 is a model of a device developed in our laboratory by L. Leifer and V. Scheinman. The idea seems to have originated with Leifer who called this device an "Orm"(*), and has since been advanced by other [7] under the name of "Tensor Arm". It is structurally a many degree of freedom manipulator with very small link lengths. Because of this it takes on the appearance of a continuous member curved in snake-like fashion. It is claimed [7] that, in use as an under-water manipulator, the Tensor Arm optimizes slenderness, cost, microdexterity, and range of operation.

The Orm model, Fig. 3, has six moving links, and is pneumatically powered. The joints between members are made up of two intersecting revolutes at 90° as in the yolk of a Hook's joint. It is possible to rotate about either of the two axes independently. Our concept is to power such a manipulator with actuators that are either full-off or full-on. Such systems would have the inherent advantage of

---

(*) "Orm" is the Norwegian word for snake; its similarity to the English "arm" is most appealing.

being digital, i.e., their actuators all have only two states: "0" or "1". Only one binary bit would then be required to control an actuator, and the manipulator position would be known from the state of the bits. This would then eliminate the need for digital-analog conversion.

In the device shown each link has eight actuators equally distributed in a circle and equidistant from the point of intersection of the two revolute axes. Thus, if there is one and only one actuator on at any given time, there are eight possible configurations. This six link model then has 8^6 configurations.

This manipulator is different not in its kinematic structure but rather due to the actuators. (an analgous, but less versatile, manipulator could be designed with stepping motors which have only eight steps.) In order for such manipulators to be useful in general applications the 8^n (assuming n-links and eight relative positions of each) configurations must fairly well cover a given working space. It has been estimated that such a 24 link system could, on the average, reach points 0.00004 inches apart in a 50,000 cubic inch volume [1].

Such manipulators seem difficult to control dynamically, but the problem is yet to be studied. The kinematic problem of placing the terminal end at a given position has been studied [1] and seems amenable to solution.

Several schemes for construction have been tested. By far the most promising uses air cylinder actuators which are fixed to the frame. Thin cables then are attached to the various links in order to transmit the actuator forces to the manipulator. The major disadvantage of the orm manipulator is that it requires 8^n actuators, valves, and cables.

Conclusions

The several studies on manipulators conducted at Stanford University under the auspices of the Artificial Intelligence project have been summarized. It is hoped that these results, and the ideas advanced herein will be useful to designers of computer controlled manipulators.

REFERENCES

[1] Pieper, D.L., "The Kinematics of Manipulators Under Computer Control." Ph. D. Thesis, Stanford University, 1968, 157 pp.

[2] Pieper, D.L., and Roth, B., "The Kinematics of Manipulators Under Computer Control", Proceedings II International Congress on the Theory of Machines and Mechanisms, 1969, Vol. 2, pp. 159-168.

[3] Kahn, M.E., "The Near-Minimum-Time Control of Open-Loop Articulated Kinematic Chains," Ph.D. Thesis, Stanford University, 1969, 171 pp.

[4] Kahn, M.E., and Roth, B. "The Near-Minimum-Time Control of Open-Loop Articulated Kinematic Chains, "Transactions of the ASME, to be published in 1971.

[5] Scheinman, V.D., "Design of Computer Controlled Manipulators," Engineers Degree Thesis, Stanford University, 1969, 53 pp.

[6] Wichman, W.M., "Use of Optical Feedback in the Computer Control of an Arm," Engineers Degree Thesis, Stanford University, 1967.

[7] Anderson, V.C., and Horn, R.C., "Tensor Arm Manipulator Design". ASME paper 67-DE-57.

[8] Dimentberg, F.M., "A General Method for the Investigation of Finite Displacements of Spatial Mechanisms and Certain Cases of Passive Joints", (English translation) Purdue Tranlation No. 436, Purdue University Libraries.

[9] Duffy, J., and Rooney, J., "On the Closures of Spatial Mechanisms", ASME paper no. 72-Mech-77 (1972).

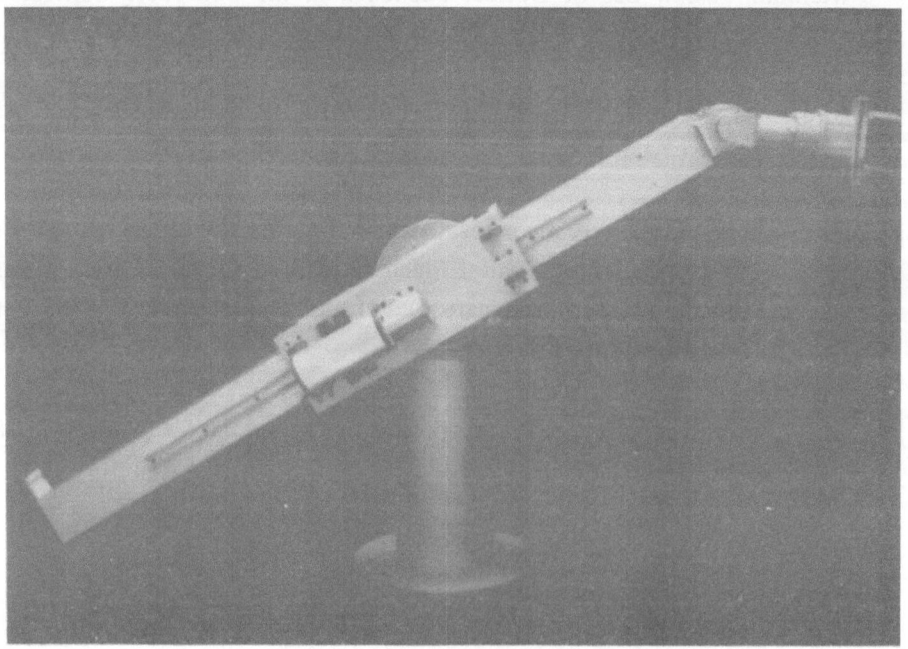

Schematic 1

Figure 1

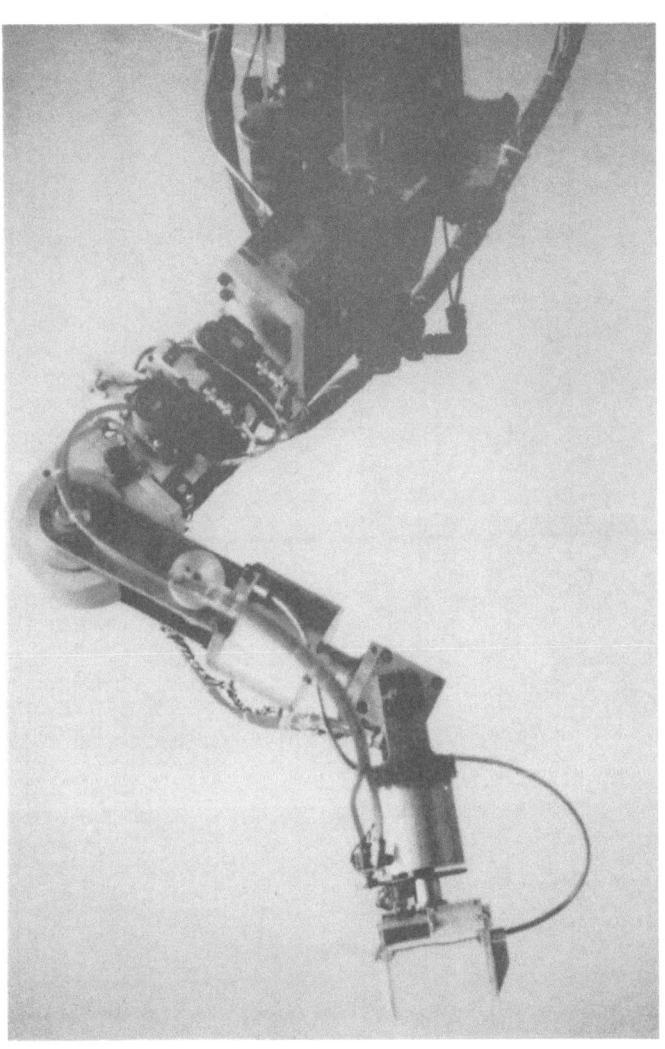

Figure 2

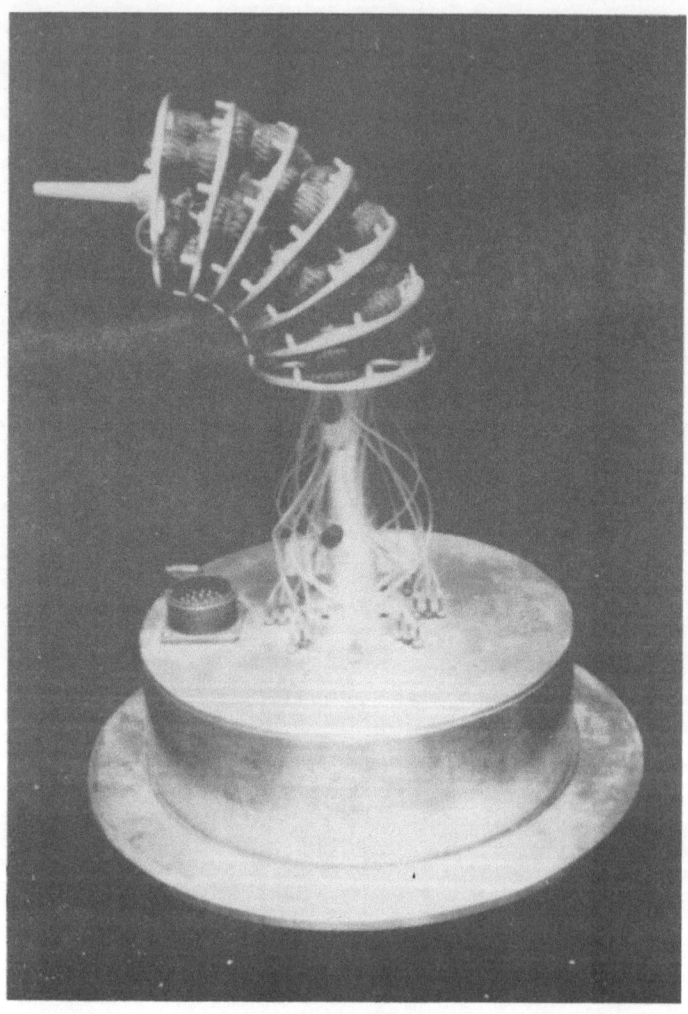

Figure 3

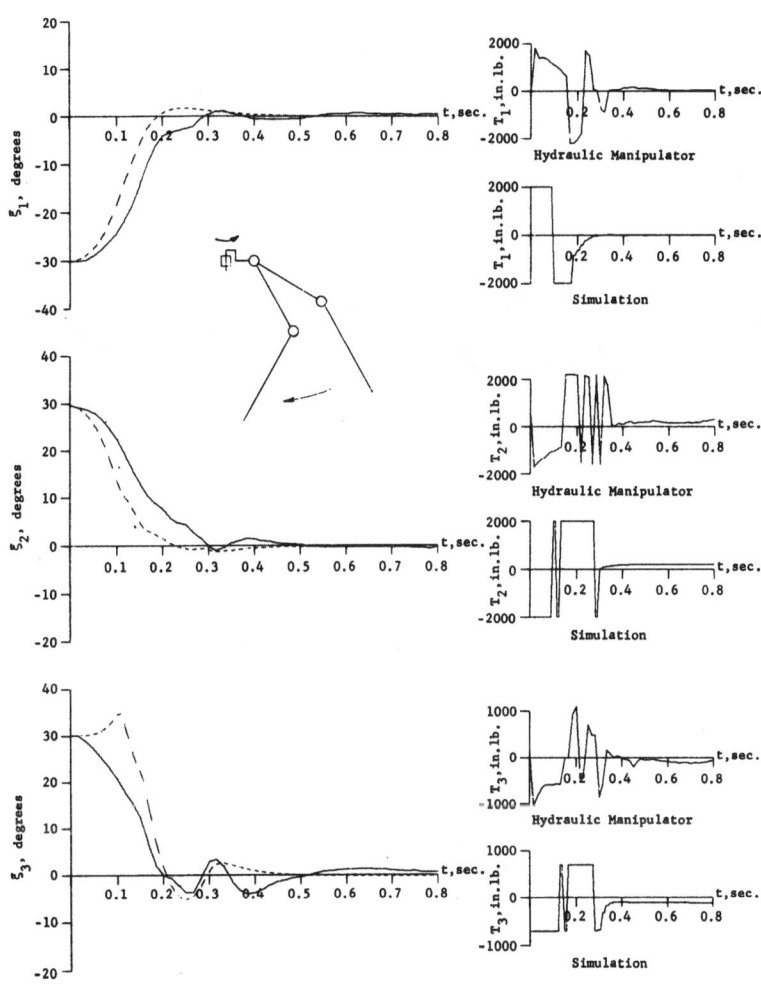

Figure 4

3. BIOMECHANICS OF MOTION

BIOMECHANICS OF WALKING UP STAIRS (*)

Aurelio CAPPOZZO,
Università di Roma,
Istituto di Fisiologia Umana,
I Cattedra di Fisiologia Umana
Roma

Tommaso LEO,
C.N.R.,
Centro di Studio dei Sistemi di Controllo
e Calcolo Automatici
Istituto di Automatica dell'Università di Roma

(**)

Summary

Some results concerning the kinematics and dynamics of walking up stairs are presented. In particular the kinematics variables are analysed by a Fourier series expansion, and the muschular moments at the lower limbs articulations are computed according to a mechanical model of the system: some energetical considerations are also included.

The results are presented in order to be used in synthesis studies and design of humanlike walking machines. The experimental and analytical method applied is a general one for the study of repetitive locomotory acts.

(*) This work was partially supported by C.N.R.
(**) All figures quoted in the text are at the end of the lecture.

1. Introduction

A deeper knowledge about the kinematics, dynamics and energetics of normal human locomotion is demanded in the design of locomotory devices, whether externally powered lower limbs prostheses, locomotory robots or other types. Such a knowledge has been adequately provided in the case of level walking. However, as far as the AA are aware, the only study related to walking up stairs is that of Morrison concerning the forces at the knee joint [1] .

The aim of this work is to supply information about the biomechanics of this locomotory act: the patterns of the kinematic variables, the prevalent components of the muscular moments acting at the articulations and the mechanical energy involved in the movement. The method applied involves experimental investigations, the results of which have been processed according to dynamic equilibrium equations implemented on a suitable mechanical model of the biological structure [10].

This work was made possible through the cooperation of the C.S.S.C.C.A. — Istituto di Automatica and of the Istituto di Fisiologia Umana of Rome University, cooperation being based on the respective scientific interests in the field of active lower limb prostheses and in the field of human locomotion.

2. Preliminary considerations

Nowadays there are some interesting studies on the human locomotory apparatus which are characterized by a multilevel hierarchical approach to the functional aspects of the system [2-3-4-5]; the levels considered are generally three: decision making, algorithmic and dynamic levels. The present work is concerned with the dynamic level.

With the typical tools of system theory, an efficient mathematical model describing such a level can be easily conceived. For a useful application of such a model to the design of some sort of artificial locomotory device, it is necessary to have, as a reference term, a synthetic description of the phenomena involved in the physiological system by stating relationships between the more significant variables.

For such a purpose an experimental and analytical method, of general use in human locomotion problems, developed by A. Pedotti and the AA., was used [10] .

Some features of this method are summarized below.

A preliminary investigation of the movement under study leads to the conclusion that the external forces acting on the structure have the outweighting components lying on the sagittal plane. In fact, through the measurement of the ground reactions, the component of the resultant of the inertia forces lying on the frontal plane, is about 25 % and 15 % of the vertical and horizontal components on the sagittal plane respectively. So it is reasonable to limit the investigation to the sagittal plane. In particular, only the lower limbs have been investigated. A multilinkage model has been adopted [6] assuming that the joint's axes do not change their distances during the motion. Under the hypothesis of complete symmetry of the gait, the analysis has been limited to a single leg. The present mechanical model is then, a three-links one with five degrees of freedom, to which the five free coordinates η_1, η_2, η_3, x_{H}, z_H (fig. 1), have been associated. From experiments on the physiological system the free coordinates of the mechanical model and the ground reactions can be obtained. Computations lead to the muscular moments and to some energetical considerations. The next step is the graphical representation of the more significant relationships between the measured and computed functions.

3. Experimental procedure

The wooden stair, on which the experiments were made, has six steps and was designed to avoid vibrations (fig. 2). The dimensions of each step were made according to a current standard (32 x 16 cm). The third step was a multicomponent force platform mounting piezoelectric transducers. The test subjects bore four markers approximating as well as possible to the rotation centres of hip joint, knee joint, ankle joint and metatarso-phalangeal joint of the left leg.

As the subject walked up the stairs a film was taken by a camera placed on his left side at a distance of 7.60 m; the frame frequency was 20 f.p.s., well defined by a flashing light generated by a stroboscope. The camera has the shutter opened during filming. Simultaneously with the film, a photo was taken of the oscilloscopic screen displaying the platform outputs. These outputs were the horizontal force acting on the plane of progression, and the two components of the vertical force whose points of application on the same plane were known from the factory's data on the platform. In this way the resultant vertical force and its instantaneous point of application (x_r) were known. The record of the platform outputs was synchronized with the film by recording the triggering pulses of the stroboscope (in the same photo).

Five male subjects, 20-30 years old, without any apparent physical deficiency, were tested. They wore their every-day shoes. Each subject was required to walk up the stairs at various speeds, chosen by himself in the range of his own natural gait. Particular attention was paid to the fact that a steady state movement was to be recorded. By projecting the film, a 1:10 image was obtained. A stick diagram was drawn, on which $x_{||}, z_H, \eta_1, \eta_2, \eta_3$ were measured.

These measures are subject to errors which have a global effect estimated at ≤ 6mm on distance, and $\leq 3°$ on angles. The measurements of the ground reactions produced an indeterminate result within ± 1.5 kg. Such errors are well absorbed by those intrinsic in the biomechanical model adopted. These problems have been taken into account in the analytical processing of the data.

4. Analytical procedure

The functions $\eta_1(t), \eta_2(t), \eta_3(t)$ are periodic with a period equal to the double step duration. $x_{||}(t)$ and $z_H(t)$ can be seen as the sum of a ramp plus a periodic function with the same period of the ones mentioned above :

$$(1) \qquad x_H(t) = K_x \cdot t + x_H^*(t) \quad ; \quad K_x = \frac{x_H(T) - x_H(0)}{T}$$

$$(2) \qquad z_H(t) = K_x \cdot t + z_H^*(t) \quad ; \quad K_z = \frac{z_H(T) - z_H(0)}{T}$$

where T = double step period.

These functions, by virtue of their periodicity, can be represented by a trigonometric series. In the hypothesis that the errors affecting the experimental data are random, the series parameters can be estimated by a least square method. For each function a number of Fourier series has been considered, each having a number of nonzero harmonics increasing from 1 to (N-1)/2 (N = number of data points). At each level of approximation the variance estimation has been calculated. Plots of $E|\sigma|$ vs. the number of nonzero harmonics (n) relative to the free coordinates and all the tests, are shown in fig. 3. The results will confirm the assumption of an estimation of the above periodic functions by Fourier series with a number of nonzero harmonics equal to $\bar{n}_i$ (see fig. 3) respectively.

The first and second derivatives of these functions have been calculated and the motion kinematics then completely described.

In the dynamics, the muscular moments M_H, M_K, M_A acting in the plane of progression on the hip, knee and ankle joints respectively, have been calculated according to the following equations :

$$M_H = J_c \ddot{\theta}_1 + m_c \ddot{x}_c (z_H - z_c) + m_c \ddot{z}_c(x_c - x_H) + m_c g (x_c - x_H) + J_g \ddot{\theta}_2 +$$

$$+ m_g \ddot{x}_g (z_H - z_g) + m_g \ddot{z}_g(x_g - x_H) + m_g g (x_g - x_H) + J_p \ddot{\theta}_3 + \qquad (3)$$

$$+ m_p \ddot{x}_p (z_H - z_p) + m_p \ddot{z}_p(x_p - x_H) + m_p g (x_p - x_H) - R_z(x_r - x_H) - R_x z_H$$

$$M_K = J_g \ddot{\theta}_2 + m_g \ddot{x}_g (z_K - z_g) + m_g \ddot{z}_g(x_g - x_K) + m_g g (x_g - x_K) + J_p \ddot{\theta}_3 +$$

$$+ m_p \ddot{x}_p (z_K - z_p) + m_p \ddot{z}_p(x_p - x_K) + m_p g (x_p - x_K) - R_z(x_r - x_K) - R_x z_K$$

$$\qquad (4)$$

$$M_A = J_p \ddot{\theta}_3 + m_p \ddot{x}_p (z_A - z_p) + m_p \ddot{z}_p(x_p - x_A) + m_p g (x_p - x_A) +$$

$$+ R_z (x_r - x_A) - R_x z_A \qquad (5)$$

The symbols can be interpreted from fig. 1.

The above equations have been obtained by applying d'Alambert's principle of dynamic equilibrium. The mechanical parameters of the structure have been evaluated according to the method described by Drillis and Contini [7] .

The mechanical power supplied by the muscular actuators at each joint can be easily calculated by multiplying the above moments and the relative angular velocities. The total mechanical energy is directly estimated by integrating the power over the double step period.

All the computations described have been implemented on the UNIVAC 1108 computex of the "Centro di Calcolo Interfacoltà" of Rome University, through a programme in Fortran V [10] .

5. Results

In Table 1 the overall characteristics of the tests from which the results have been obtained, are summarized. The whole stride has been divided in two phases: stance, in which the foot is in contact with the ground, and swing. The average stance phase duration is 64 % of the stride duration [8] .

Kinematics: the first result concerns the distribution of modulus and phase of each harmonic of η_1, η_2, η_3 (fig. 4). The plots exhibit a good thickening of the values for all the tests as far as the $\bar{n}_i$-th harmonic. For $n > \bar{n}_i$ the phase suddenly assumes values scattered at random between $-\pi$ and $+\pi$, while the modulus becomes definitively negligible. This fact states that the gait is not characterized by harmonics of an order greater than $\bar{n}_i$. The values relative to η_3 have a great dispersion, which must be correlated to the individual mode of each subject of placing the foot on the step.

It must be emphasized that a probabilistic description of the above distribution has not been attempted because of the inadequate number of tests.

About the analogous plots relative to x_H^* and z_{II}^* it must be observed that a satisfactory matching is exhibited only by those tests at different speeds of the same subject. In all the tests and for both the functions, first and second harmonics have modulus values lower than about 30 mm, while the other harmonics have negligible modulus values. While the measurements percentage errors are remarkable, it should be emphasized that a first harmonic which takes into account the transverse and frontal rotation of the pelvis does exist.

An example of plots vs. time of the free coordinates is shown in fig. 5.

For a synthetic representation of gaits the plot η_1 vs. η_2 is often used in the literature [2-3-8]. For the locomotory act under consideration such a plot is shown in fig. 6, for one subject at various speeds. Such a diagram exhibits a good invariance with respect to the speed and to the test subject. Only in tests 3 and 13 did the curve undergo characteristic changes of shape, these changes consisting in a top part with η_1 quite constant and in a reduction of both angle ranges.

The phase plane plots of the three angular coordinates are shown respectively in figs. 7, 8, 9. All these plots are readily repeatable, with the exception of the two tests with a stride period far below the second. In these exceptions the most marked differences are observed in the plots concerning η_2.

Dynamics: in fig. 10 an example of ground reactions is shown. It is emphasized that in steady-state conditions the mean value of R_x must be zero, and the mean value of R_z must be equal to the body weight. During computation this property has been used to verify the correspondence between the experimental situation and the a priori hypothesis of periodicity of the free coordinates functions.

An example of muscular moments at the joints is plotted in fig. 11. It must be pointed out that the hip moment M_H suffers the cumulative effect of all the errors. Despite this, some general characteristics of M_H are evident : the stance phase can be subdivided into two parts, the first having extensorial torques prevalent, the second flexorial ones. In the swing phase a flexorial action is always followed by an extensorial one.

The knee moment M_K has quite good repetition in the cases examined. During the support phase an extensorial torque is always strikingly prevalent.

Relatively small torques, extensorial and then flexorial, characterize the swing phase.

The ankle moment M_A has a significant value only during the stance. Its typical two-peaks shape is strictly recurrent; its only variations are in the relative values of the two peaks. The torque during the swing phase is negligible. The inertial and gravitational components of the above torques decrease from the hip to the ankle, where they are negligible.

In Table 2 the features of the three moments during the stance phase are quantitatively summarized.

In figs 12 and 13, the plots moment vs. angular velocity for the hip and knee joint respectively, are shown. A similar plot for the ankle joint is not given, since the complexity of its geometry conceals any significant information. Such plots synthesize what has been said about the dynamics and kinematics of the motion, and allow a direct inspection of the maximum powers involved, either absorbed or supplied by the actuators.

Energetics: the mechanical energy increase (E) associated with the lower limb during a double step is shown in Table 3 with the maximum and mean values of the power related to each joint. This increase, evaluated for both the limbs, amounts to 70-90 % of the total body energy increase ($V-V_0$) during a stride, which coincides, according to the periodicity of the motion, with the potential energy increase associated with the body mass centre. This result is most probable and can be considered a good check for all the above results.

The values of the joint's powers suffer from the individual modes of walking up stairs as well as from the experimental errors. A scattering of the power contribution at each joint can be observed, even among tests of the same subject. A reliable hypothesis on this fact can not be proposed at this state of the research.

6. Conclusions

By a finite order Fourier series a mathematical description of the gait kinematics was obtained. This description, characterized by a limited number of parameters, allowed the analytically exact computation of the dynamics and energetics functions.

The most significant results obtained using this method may be summarized as follows. A characteristic locomotion pattern in walking up stairs exists for all gaits with a stride period from 1 to 1.50 seconds. In this speed range a correlation between the antropomorphic parameters of the subject and his locomotion pattern cannot be proved, at least, not in the relatively small number of tests carried out. When the stride period is out of this range some variations are evident, which could be correlated to a change of gait. This hypothesis must be verified by an adequate number of tests.

The set of kinematics, dynamics and energetics characteristics obtained for walking up stairs constitutes features for the design of active prostheses and is an useful reference term for the synthesis of more sophisticated biped locomotion machines.

The results presented in this paper also have a physiological relevance. From a physiological point of view a deeper investigation with a larger number of tests seems to be called for. For this purpose an automization of the experimental procedure is imperative. This automization could be obtained by:
1) – automatic acquisition of the platform outputs
2) – automatic reading of the film recording the movement.
The AA. intend to procede along this line.

AKNOWLEDGEMENTS

The AA. wish to thank Mr. C. Iadecola, Mr. G. Federico and Mr. S. Medici for their assistence in gathering the test's data, and students and staff of the Institutes with whom the experiments were made.

REFERENCES

[1]   MORRISON J.B., "Function of the Knee Joint in various Activities", Bio-Medical Engineering, v. 4, n. 12, dec. 1969, pp. 573-580.

[2]   CHOW C.K., JACOBSON D.H., "Studies of Human Locomotion via Optimal Programming", Mathematical Biosciences, v. 10, 1971, pp. 239-306.

[3]   VUKOBRATOVIC M., JURICIC D., "Contribution to the Synthesis of Biped Gait", IEEE Trans. on Bio-Medical Engineering, v. 16, n. 1, Jan 1969, pp. 1-6.

[4]   PEDOTTI A., CAPOZZO A., GILARDI L., "Analisi del coordinamento muscolare nella deambulazione", I Convegno-Mostra di Bioingegneria, Milano, June 1972.

[5]   BERTUZZI A., LEO T., MONTELEONE M., VITELLI R., "Improvements of Human Locomotion by Powered Lower Limb Prostheses", 4-th Int. Symp. on External Control of Human Extremities. Dubrovnik 28/8-4/9, 1972

[6]   BRESLER B., FRANKEL J.P., "The Forces and Moments in the Leg during Level Walking", ASME Trans., Jan 1950, pp. 27-36.

[7]   DRILLIS R., CONTINI R., "Body Segment Parameters", Tech. Rep. 1166.03 Res. Div., School of Engng.; New York University, Sept. 1966.

[8]   MURRAY M.P. et al., "Walking Patterns of Normal Men", J. Bone Joint Surgery, v. 46-A, , n.2, Marc 1964, pp. 335-360.

[9]   PAUL J.P., "Forces Transmitted by Joints in the Human Body", Proc. Instn. Mechanical Engrs., v. 181, pt. 3J, 1966-67, pp. 7-15.

[10] CAPOZZO A., LEO T., PEDOTTI A., "General computing method for analysis of human locomotion", to be published.

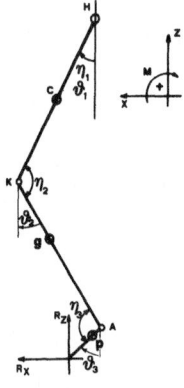

fig. 1

fig. 2

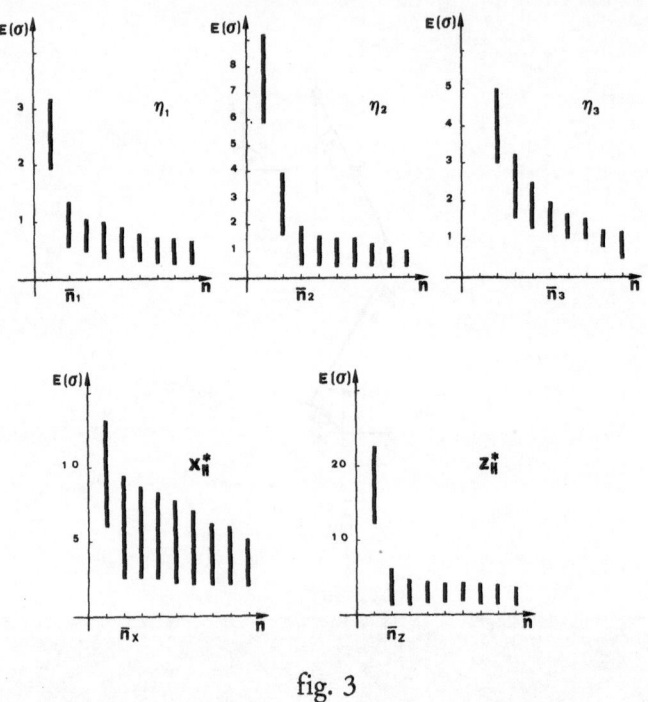

fig. 3

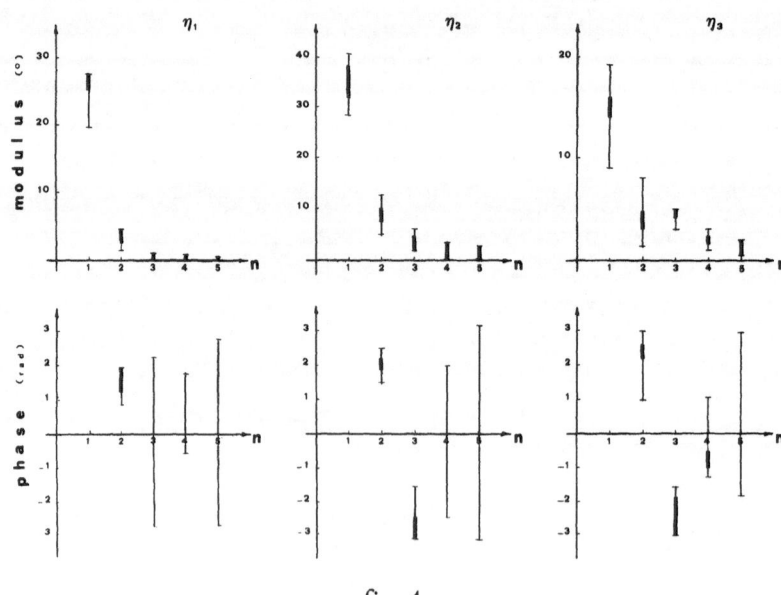

fig. 4

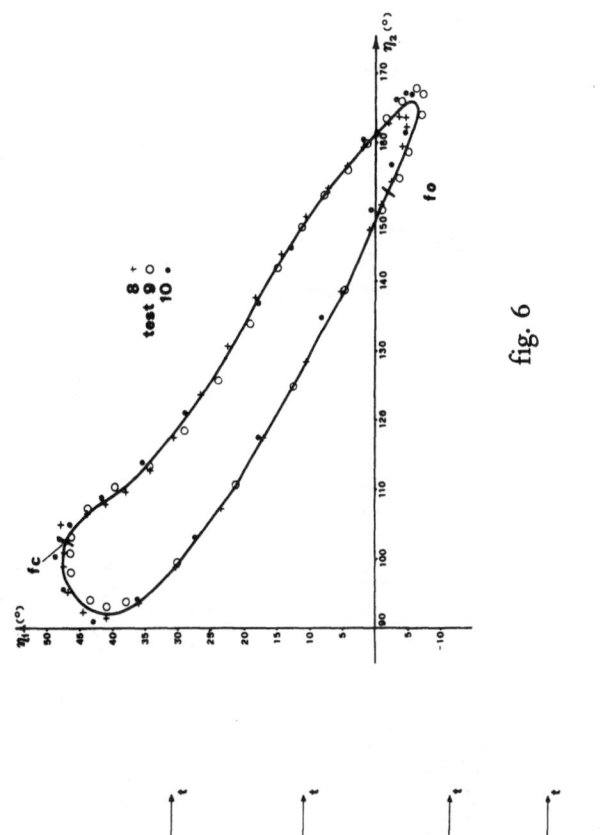

fig. 6

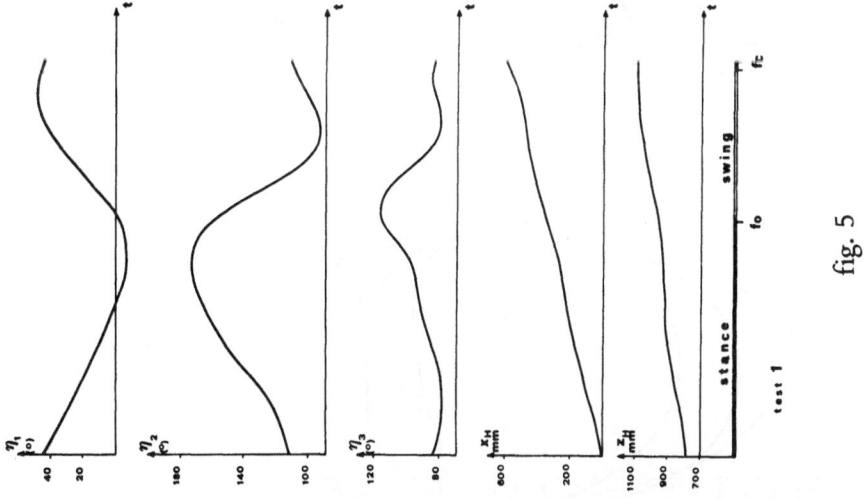

fig. 5

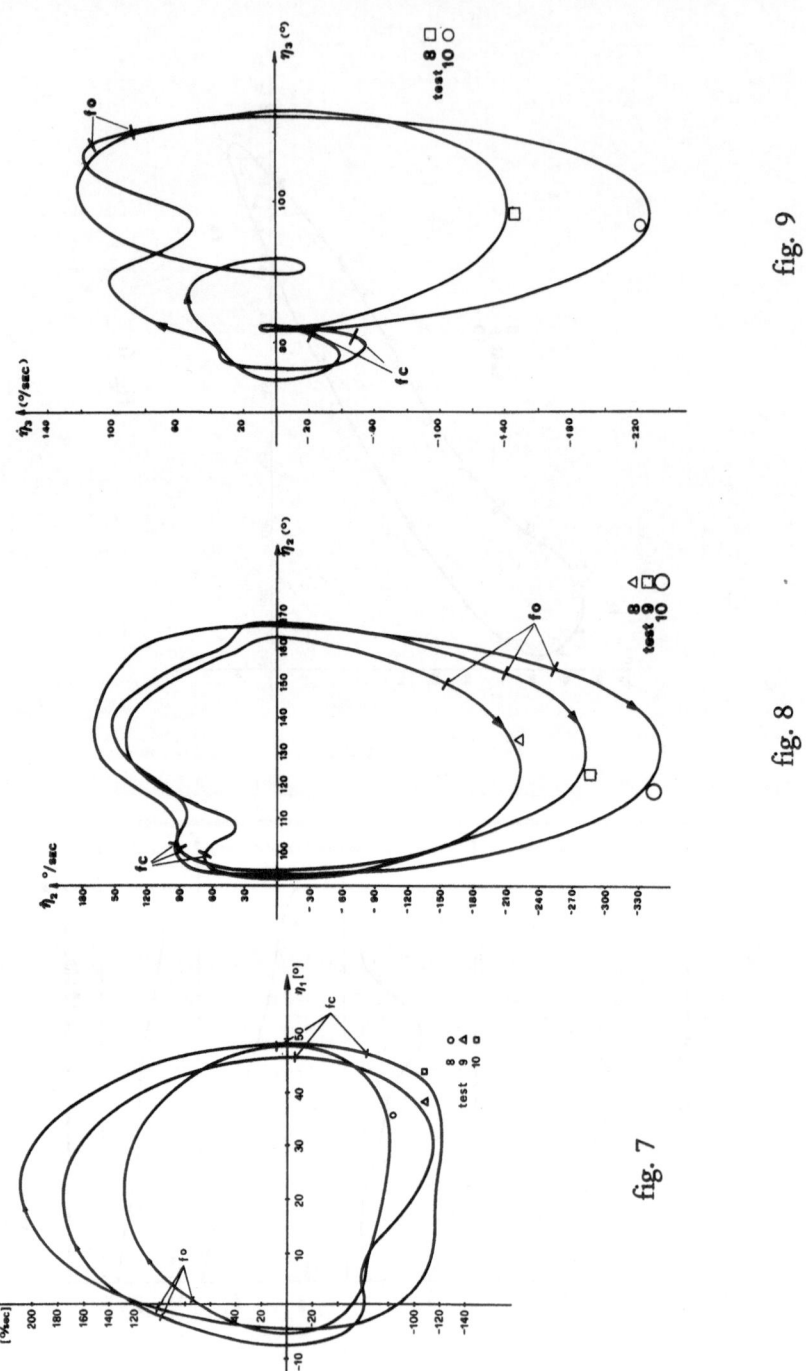

fig. 9

fig. 8

fig. 7

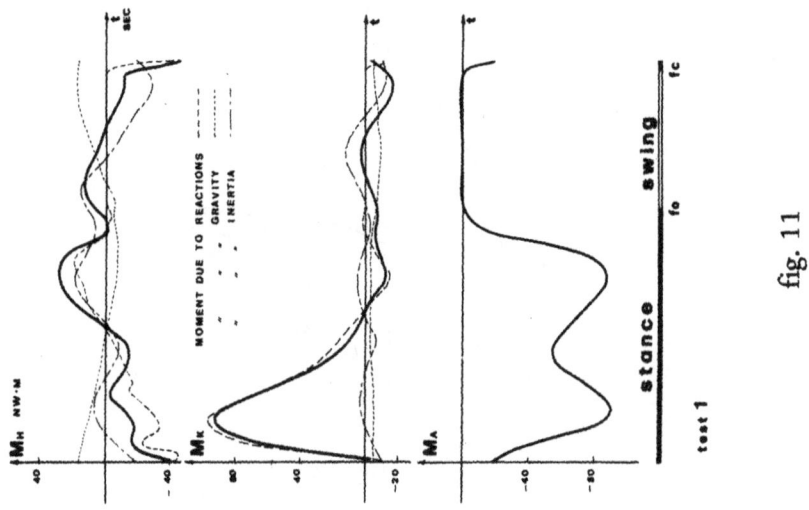

fig. 11

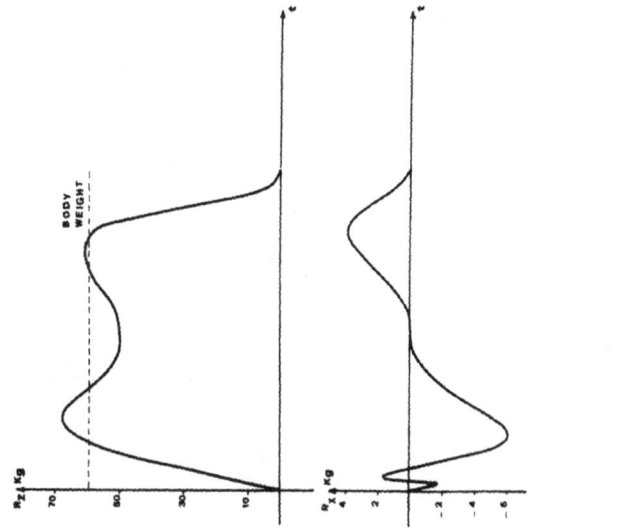

fig. 10

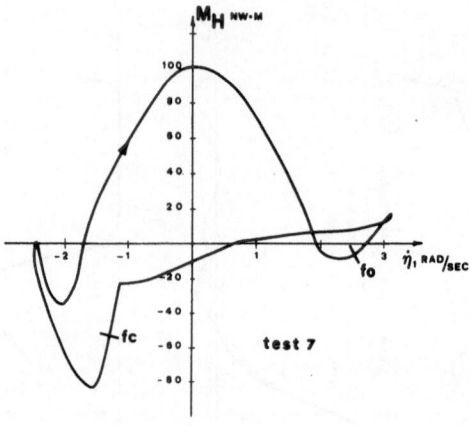

fig. 12

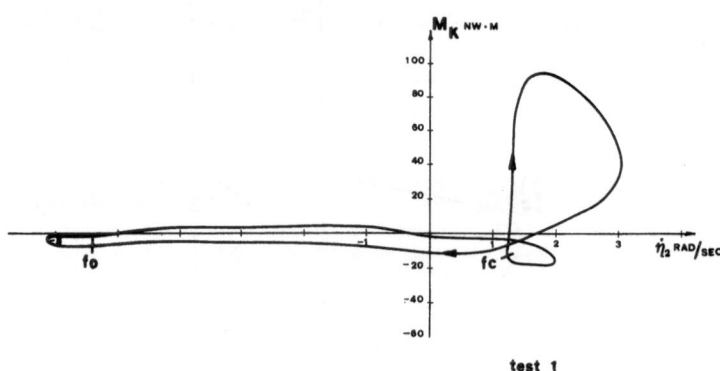

fig. 13

TABLE 2

TEST NUMBER	FLEX. MAX.TORQUE (NM.M)	FLEX STANCE %	EXT. MAX.TORQUE (NM.M)	EXT STANCE %	FLEX MAX.TORQUE (NM.M)	FLEX STANCE %	EXT MAX.TORQUE (NM.M)	EXT STANCE %	EXT MAX.TORQUE (NM.M)
1	30.0	0.45	44.0	0.55	13.0	0.41	94.0	0.59	92
2	54.0	0.36	91.0	0.64	18.0	0.25	90.0	0.75	94.6
3	65.4	0.42	124.0	0.58	10.4	0.21	107.3	0.79	97
4	13.0	0.62	45.3	0.38	30.6	0.45	87.0	0.55	112.1
5	37.0	0.74	87.6	0.26	26.3	0.44	129.6	0.56	125.3
6	30.0	0.38	35.0	0.62	22.8	0.41	104.4	0.59	100
7	101.3	0.4	84.9	0.6	10.0	0.16	97.5	0.84	95.6
8									
9	1.2	0.14	63.8	0.86	19.4	0.5	70.7	0.5	91.6
10	19.7	0.13	65.0	0.87	25.2	0.22	66.1	0.78	86.5
11	36.7	0.38	87.7	0.62	9.7	0.43	94.8	0.57	114.4
12	32.0	0.4	88.7	0.6	10.7	0.34	104.0	0.66	113.6
13	17.0	0.33	193.6	0.67	22.1	0.38	103.0	0.62	165.2

TABLE 1

SUBJECT	AGE	WEIGHT (KG)	HEIGHT (M)	NUMBER	STRIDE DURATION (SEC)	STANCE DURATION (SEC)	STANCE/STRIDE %
GF	26	63	1.69	1	1.15	0.75	0.650
				2	0.95	0.625	0.657
				3	0.80	0.525	0.657
SM	24	82.7	1.81	4	1.75	1.100	0.628
				5	1.05	0.625	0.595
WM	22	67.2	1.75	6	1.50	0.950	0.632
				7	1.00	0.625	0.625
FL	21	59	1.69	8	1.65	1.125	0.680
				9	1.30	0.900	0.692
				10	1.00	0.600	0.600
MF	30	80	1.81	11	1.40	0.925	0.661
				12	1.35	0.810	0.600
				13	0.85	0.525	0.618

TEST NUMBER	MEDIUM POWER			POWER SUPPLIED MAXIMUM			POWER ABSORBED MAXIMUM			E ONE LIMB	E TWO LIMBS	$V - V°$	Δ
	H	K	A	H	K	A	H	K	A				
1	12	34.0	30.0	71	208	183	28	31	39	87.3	174.6	203.6	29
2	16.5	39.0	38.0	153	190	263	77	63	24	88.8	177.6	203.6	26
3	18.5	45	35.0	225	260	250	102	96	30	89.1	178.2	203.6	25
4	11	28	20	60.0	151.0	197.0	17.0	33.0	41.0	103.25	206.5	258.0	52.5
5	7	49.5	23.5	113.5	324.0	204.0	64.0	44.0	51.0	84.05	168.10	258.0	89.9
6	17.0	29.0	12.0	50.0	180	127	17.0	36.0	23	87	174	211.0	37.0
7	30	27	25.0	133.0	323.0	285.0	50.0	81.0	17.0	82	164	211.0	47.0
8													
9	25.5	25.0	18.0	91.0	119.0	84.5	6.0	15	52.0	89.1	178.2	185.0	6.8
10	22.5	37	28.0	71.0	158	108.0	11.0	40.0	19.0	87.5	175.0	185.0	10.0
11	27	41.5	14.0	105	210	103	21	30	64	115.4	230.8	251.0	20.2
12	29	35.5	11.0	121	189	79.5	28	29	69.5	102.0	204.0	251.0	47.0
13	58.0	25.0	44.0	235	211	238	21	89	109	108.0	216.0	251.0	35.0

TABLE 3

BIOMECHANICAL PRINCIPLES OF CONSTRUCTING ARTIFICIAL WALKING SYSTEMS

Victor S. GURFINKEL, Professor,
IPIT Acad. Sci. USSR, Moscow, USSR

Sergei V. FOMIN, Professor,
IPIT Acad. Sci. USSR, Moscow, USSR

(*)

Summary

Constructing walking systems is an essential aspect of general problem of design of various robot-like mechanisms. Locomotor system of the vertebrates is a natural prototype of such systems.

Walking systems of man and animals have many essential differences from "standard" technical devices. These differences involve their structure, kinematics, dynamics and control. The peculiarities of living locomotor systems and an attempt to implement the ideas developed through the analysis of locomotor systems in an artificial design are the subject of this paper.

(*) All figures quoted in the text are at the end of the lecture

1. Introduction

Within the scope of general problems related to the development and design of various robot-like mechanisms a special place belongs to the problem of constructing walking systems. This is determined both by the prospects of practical applications of such systems and also by the possibility of discussing in this connection a number of general problems related to the construction of automatic mechanisms of various types (manipulators, general purpose robots, etc.).

It is but natural that in developing a walking system we turn to the locomotor apparatus of the vertebrates as the biological prototype of any walking system.

The walking mechanisms of man have been studied most thoroughly, but our knowledge of the kinematics and dynamics of walking is scarce even to-day. Nevertheless, we already know enough to attempt some generalizations which may prove useful for designing artificial locomotor systems.

Essential differences are apparent even on a superficial comparison of the "standard" technical devices with living things and, particularly, with locomotor systems of man and animals.

In technical devices the qualities of a system as a whole (reliability, performance rate, accuracy) are determined by its weakest links and are, as a rule, inferior to the corresponding qualities of its parts. On the contrary, the quality of a living system as a whole is superior to the corresponding qualities of its elements. For example, living systems are capable to go on with their function being damaged even essentially, though the reliability of their elements (cells) is but small. Living systems which are built up of "slow" memory elements are more efficient than computers at solving some problems, e.g. problems of recognition.

Finally, in living systems there is a mechanism of mutual error correcting in different parts (which is most important for problems of locomotion) so that the resulting overall accuracy of the system is greater than that of its parts.

2. Structural properties of the locomotor system

During the locomotion there is a periodical change of the structure of the locomotor system. Its links form a closed kinematic chain in two-support phase and an open chain in single-support phase. The links and kinematic pairs of locomotor system have some important peculiarities. Thus, the bulk of extremities consists of muscular tissue ; the partial gravity centers of the links are shifted in a proximal direction. This peculiarity is connected with the fact that the accelerations in the distal part of the leg are much greater than in the proximal part.

The joints represent kinematic pairs of different classes and the amplitudes of their movement are different. It is worth noting motion in joint is restricted by elastic structures and not by rigid stops. Every degree of freedom in legs is ensured by several muscles, and not by just one, i.e. the number of muscles is much more numerous than the degrees of freedom.

3. Kinematic properties of the locomotor system

The leg is a hinge system in the following two basic sets of conditions: longitudinal compression (during the stance phase). The requirements that the leg should meet to satisfy these two states are mutually contradictory in their essence, and the problem of combining reliability of support and the ease of swing is hard to solve, all the more so that in many cases motion occurs on an uneven surface. Adequate vertical adaptation and absorption of the dynamic loads during the support phase are ensured by the fact that the inter-link angles in the supporting extremity are as a rule different from zero, which means that during the stance phase as well as in other situations an extremity does not stretch completely.

During the swing phase the leg can be treated as a freely suspended pendulum. This is confirmed by the fact that changes of the inter-link angles in leg joints during walking are described by functions closely resembling simple sinusoidal waves.

As has been confirmed by numerous observations, the relative positions of the links in an extremity as well as their positions with respect to the body are never repeated during the motor cycle (two step cycle). In other words, the configuration of the motor system at each moment of time uniquely determines the phase of motion and therefore the subsequent phases of walking.

Quite often human walking as a first approximation has been treated as a planar motion. However, many significant details of true walking are lost in this case, since some of the small displacements (including those in the plane perpendicular to the direction of walking) are very essential for locomotion. The significance of the small amplitude displacements is apparently connected with the large values of the amplification coefficients in the appropriate feedback loops. Experimental attempts to set up artificial human planar walking in systems of the exoskeleton type resulted, for reasons just mentioned, in a strained, unnatural and energetically wasteful gait.

4. Dynamic aspects of locomotion

Dynamics of the locomotor system of the vertebrates, and in particular, of the locomotor system of man, is ensured by peculiar engines, the muscles. Peculiarity of these chemomechanical engines stems not only from the processes of conversion of the chemical energy into the mechanical energy which occur there ; the muscle macrostructure is also of great importance. Firstly, the entity called a muscle in anatomy is, from a functional standpoint, a set containing a great number of jointly working (but not in a synchronous way) engines, which are somewhat different with respect to their mechanical characteristics. Secondly, a muscle is an engine in which the contractive part is always connected in series with some elastic element. And, finally, an important property of any muscle is strong non-linearity of its characteristics.

Muscle activity in locomotion is not constant : muscles are activated during the periods of acceleration and deceleration only. The result of their action is a series of pulses, which are much shorter than the locomotive cycle.

There exists a certain rate of walking which is energetically optimal (for man it corresponds to a speed of approximately 3.5 km/hr). It occurs when the extremities during the swing phase move in the conditions of free oscillations. Energy expenditures are needed to change the speed of walking. However, the presence of non-linearities in the connective elements between the links of the locomotor system as well as the possibility to control rigidity of these connective elements makes it possible to increase the rate of walking considerably without rapid grow of the energy expenditures.

In contrast with kinematics of walking, which can readily be observed, such dynamic characteristics of walking as momenta appearing in joints due to muscle action are hard to be registered directly. However, they can be calculated on the basis of the equations of motion of the hinge system. If the kinematics and reactions of the support are known.

Such calculations have recently been done in our Institute. They confirm the qualitative results which were known earlier: relatively short periods of high muscle activity versus the whole locomotive cycle, strong correlation of the momenta acting in various joint, etc. Besides, considerable step--to-step variability of momenta has been foundç which is essentially larger than the apparent (kinematic) deviations of walking from the strictly periodic pattern.

5. Control

The idea of a multilevel organization of control in the locomotor system has originally been formulated and discussed by N.A. Bernstein. The basic principles of this organization are as follows :

1) There is a central (top level) program of movements.
2) This program is realized through the periferal motor and sensory mechanisms.
3) Besides the inter-level "vertical" interaction (which is displayed in a higher level influence upon the lower one and vice versa) an important part belongs to the interaction of the elements within the same level, the so-called "horizontal" interaction.
4) Various types of inter-level interaction exist, in particular :

a) action coming from a higher level aimed at rearranging the interaction between the elements of the lower level ;

b) modulation of the global functional activity of the lower level ;

c) direct control "from the top" over the elements of a lower level.

5) Every level is autonomous within wide limits ; it is capable of "independent action". This is expressed in that the scope of actions that can be taken at each level is much larger than the set of instructions coming to the level "from above".

It can be said that in biological systems each particular problem is solved, as a rule, on the lowest possible level from among those where it can be solved at all ; higher levels are relieved, as much as possible, of "secondary" duties.

6) The autonomous behaviour of the lower levels in locomotor systems is combined with strictly hierarchical organization : only those actions can be taken on a lower level which don't contradict the instructions coming from higher levels.

A good illustration of these general principles is given by the preparation of a mesencephalic cat, which was developed a few years ago by M.L. Shick. Such a cat cannot walk unassisted, and when placed on a moving treadmill remains completely passive.

However, the cat can be forced to walk by electric stimulation (or to run, if the stimuli are strong enough), with coordination of all types of movements being practically normal.

This means that the main part of the locomotor program is realized by the lower levels of the nervous system and not by the top ones. In this case only a general instruction determining the intensity of movements comes from above.

6. Stability

Control over the locomotor system consists not only of ensuring the movement, because walking is not just a process of flexing and extending the extremities by preserving equilibrium of the body. This is why mechanisms of body postural control are of considerable interest and are equally important as the motor program.

The following points should be kept in mind in connection with this :

1) posture preservation (postural activity) is a very usual task for the locomotor system ; any movement starts from some posture and ends in another posture, which may be the same or different from the original one.

2) many dozens of muscles take part in realization of postural activity, much more than in performing a "macromovement".

3) postural activity requires a much greater accuracy of control than that needed for "macromovements".

All these statements are meaningful for locomotion as well, since the task of keeping equilibrium is always associated with the problem of motion.

In this connection it seems relevant to define explicitly what is meant by stability of a locomotor system.

Static stability, i.e. keeping the vertical projection of the center of gravity within the support area, does not take place over a considerable part of the locomotor cycle not only for human biped walking but for quadruped walking of the vertebrates as well. This is why locomotor problems should be discussed in terms of dynamic stability of the system.

The adequate definition cannot be given in terms of the classical stability theory since a locomotor system has considerable freedom in selecting the control modes. As a consequence, no state of the system at $t = t_0$ can guarantee equilibrium at $t > t_0$ .Therefore, equilibrium of a system which uses legs for walking should be treated only as its ability not to fall (under proper control) at consecutive moments of time. To be more precise, let X denote the phase space of our system, and x_1 , x_2 some of its two states. We say that the state x_2 is accessible from the state x_1 if there exists a mode of control (from the set of those available for the system) which lets the system pass from x_1 to x_2 during a certain interval of time. The aggregate of mutually accessible states, including the position of rest, will be called the reversibility domain of the system. We say that the system preserves (dynamic) stability if it stays within the reversibility domain indefinitely long.

It follows from this definition that stability is an active process for which control algorithms are vital.

As a rule, keeping equilibrium in locomotor tasks is not an end in itself. But solution of the main task, say the task of motion in a given direction, is to be paralleled by keeping equilibrium. Moreover, it there appears a danger of losing stability (the system involved approaches the boundary of the reversibility domain), preservation of the stability of the system becomes the task of first priority dictating selection of a proper control mode.

7. Body schemata and the role of receptors

An important role in ensuring stability belongs to the "body schemata", i.e. the ability of an organism to determine the mutual positions of the elements of the body. This ability develops in an organism in the process of its individual development. The mechanism of formation of the body schemata is not quite clear. Helmyolz attributed to vision an important role in it, Sechenov emphasized the significance of proprioreception. In accordance with I. Zyon the vestibular apparatus plays a leading role in the formation of the body schemata and development of spatial orientation. Anyway, it is clear that interaction of all afferent systems is essential both for the formation of the body schemata and for current determination of the position of the body.

Locomotor system of the higher vertebrates is rich in receptors that can supply information on muscles lengths, rates of change of these lengths, the forces involved, position of the links in a joint, the magnitude, rate and direction of change of the mutual position of these links, the sensitivity of some of these receptors being subject to control.

Reception in joints apparently plays the most important role in realization of the body schemata and, on that basis, in ensuring stability. The matter is that it is hard to get information about the spatial positions or the links of the body from the data on muscle lengths and the rates of their change. The reasons for this are as follows. First of all, as a rule, each movement involves many muscles; secondly, a considerable portion of the muscles belong to the two-joint type, and therefore their state does not determine the positions of links in one joint ; thirdly, information coming from the muscle length receptors depends not only on muscle length but on the way these receptors are tuned as well. Each of the joint receptors, as a rule, works within a certain range of angles ; the set of all these ranges includes all possible values of the joint angle.

In recent years interesting data have been obtained on the central control of receptors. It has been shown, for example, that certain types of interaction, which are effected through central structures, exist between visual, eye-moving and vestibular receptors.

Interaction of this type can either amplify or block up signals coming from various receptors during different phases of motion. This type of receptor interaction may be of interest from the viewpoint of ensuring stability of the artificial systems.

8. Some approaches to the design

We have made an attempt to make use of the ideas extracted from the analysis of locomotor systems in designing a moving structure, which is yet far from being perfect.

The structure is a small four-wheel car provided with a pair of spider-like legs. The wheels serve as a passive support. The distal parts of the legs carry sensors, so that the legs can "feel" the supporting surface and come down to the proper level when the car is placed on a support. Each leg has 3 joints, i.e. there are 6 degrees of freedom in all. On each leg there is a leading joint, which normally determines the work of the two others. There is also a certain degree of coordination in the performance of both extremities. At the same time, there is a possibility of direct control from the highest level over each of the elements of a lower level. This enables the car to make turns, change the velocity of movement or overcome some obstacles.

With reference to the structures of the type we are considering, the question arises of the most reasonable distribution of control functions between the highest level and lower levels. The highest level is most naturally realized either through electronic computers or through direct human control. We believe it was worth while to realize the main part of the locomotor program (synergy of joints, interextremity synergy, etc.), autonomously as special-purpose units mounted on the car itself (Fig. 10 represents schematically the organization of control).

Standard quasi-neuron cells having several activating and inhibiting inputs have been selected as the elements to use for the construction of appropriate units (in accordance with early results of McCulloch and Pitts any logical functions can, in principle, be realized with such elements). The autonomous control is organized in such a way that in case of direct intervention from the highest level some of the elements used in autonomous control also serve for the "direct" control. This situation is analogous to the one observed in the living systems, where, for example, the same mononeurons activate the motor units both in object-oriented and in reflex-induced types of movement.

An important shortcoming of the structure we have just described is its inadequate provision with receptors and the resulting absence of the "body schemata". As analyses of the proper biological prototypes show, development of a system which would, even in a primitive form, make it possible to fell the "body schemata" and mutual arrangement of the links is a difficult problem. Its solution, however, cannot be avoided if we wish to design a walking system capable of sufficiency complicated locomotor behaviour.

ON ANALYSIS AND SYNTHESIS OF DISTRIBUTION
OF DRIVES IN ALIVE MANIPULATORS AND PEDIPULATORS

Adam MORECKI, Professor
Institute for Applied Mechanics
Warsaw Technical University
Warsaw, Poland

Kazimierz FIDELUS, Doc. dr. hab.
Institute for Biologic Sciences
Academy of Physical Education
Warsaw, Poland

(*)

Summary

This paper deals with an analysis of distributions of muscles and their activities in different extremities of six mammals and a bird.

On a basis of this analysis a general procedure for synthesis of extremity-like manipulators and pedipulators is suggested.

(*) All figures quoted in the text are at the end of the lecture.

1. Introduction

Methods and results of investigations of the structure and function of bio-kinematik chains of muscle drives for a human and a horse are given in [1, 2 and 3]. This paper gives a wide analysis of performance and distribution of muscles which has been carried out on an example of several various upper and lower extremities of mammals and birds. The results obtained from the analysis enabled us to make some generalizations which may set up a basis for synthesis of extremity-like manipulators and pedipulators with multiple-degrees-of-freedom and various numbers of activities.

The synthesis works on the assumption that the distribution of muscle drives in a living organism is optimal from the standpoint of energy consumption and time of action.

2. Initial assumptions adopted for structural and functional analysis of drives of mammals and birds

2.1 Basic definitions

The following basic definitions have been adapted for analysis:
— muscle acton is a part of a whole muscle, whose fibres are so placed that they can develop, a force in a definite direction, or a moment of a force about the axes of a joint
— class of the muscle acton is defined as the number of joints over which it passes and develops moments of forces about the axes of these joints,
— function (activity) of the acton may be of the following six types: flexion, extension, pronation, suppination, abduction or adduction,
— number of functions (activities) of the acton is obtained by adding up components of moments of forces developed by muscle actions about axes of all the joints in which they act. This way of expressing the number of functions (activities) of the action is justified by the principle of operation of a control system in the organism, which simultaneously sends signals to motor units engaged in given moment.

The assumed definitions and methods of schematization adopted in the theory of structure of mechanisms with rigid links, together with the results of wide anatomical functional studies, served as a basis for a numerical comparison of muscle actons and their functions for 12 upper and lower extremities of some mammals and a bird [6, 7, 8].

The types of the extremities under discussion are a sufficient sample on the basis of which to make generalizations. What makes the sample sufficient is the number of degrees of freedom, and the number of functions and classes involved.

Table 1 shows the numerical results of the comparison made between functions and acton classes for one of the simplest structures, namely a foreleg of a horse, having w = 9 degrees of freedom, 29 muscle actons and 72 activities.

The latin names of actons, the activities in individual joints, the class of acton and the number of all activities in the joints, are given in columns 2, 3, 4 and 5 of the table, respectively.

2.2 Method of drawing histograms for distributions of functions and classes of muscle actons

The purpose of this procedure is to supply information about the distribution of functions and classes of muscle actons for individual joints.

The numbers of various activities of muscle actons for individual joints of a foreleg of a horse together with the histogram of their distribution are given in Figs 1a and 1b, respectively. It can be seen from column 1 of table 1, that among 15 muscle actons servicing the shoulder joint there are actons with 1, 2 and 3 different activities (column 5, table 1). The elbow joint (items 13-26, table 1) is serviced by actons with 1, 2, 3, 5 and 6 different activities. Joints 3 and 4 have actons with four different activities, joint 5 with four and joints 6 and 7 with two different activities. On the basis of this analysis a histogram may by drawn (Fig. 1b), from which it is clearly seen that 2, 3 and 4 different activities of one acton occur for six joints, and 5 different activities occur in one joint only.

An analysis in terms of classes of actons given us quite different results (Fig. 1c and 1d). It is evident from this analysis that actons of two different classes appear in 4 joints actons, of four classes appear in 2 joints, and actons of five different classes appear in 1 joint. There is no case with 3 different classes.

The results of such an analysis for various structures of upper extremities of 6 mammals and a bird, are given in Fig. 2. This figure presents the histograms of the distribution of the number of various classes of actons for a foreleg of a horse (b_1), cow (c_1), hen-wing (d_1), pig (e_1), rabbit and dog (f_1) and human (g_1). For comparison an example of histogram for a simple extremity-like manipulator [4] is given in Fig. $2a_1$.

Histograms of the distribution of the numbers of the various activities of actons in joints are given in Figs $2b_2$ through $2g_2$. The corresponding structural schemes for the extremities under analysis with mobilities from $w = 9$ to $w = 30$, are given in Figs $2a_3$ through $2g_3$. The mobility of the manipulator is $w = 7$. It should be noted, that these structural schemes of forelegs have an almost constant number of links and kinematic pairs of the connecting members, and a gradually increasing number of fingers of the ending tip, from 1 to 5. The histograms shown in Fig. 2 have different shapes. They may be skewed as in Fig. $2f_1$, $2e_2$, $2f_2$ or symmetric as in Figs $2e_1$, $2g_1$, $2g_2$. The histograms may be skewed to the right (positively), for example Fig. $2e_1$, or to the left (negatively) as in Figs. $2g_1$, $2g_2$. They also may be flattened to a different degree. Using the widely-received measurements of skewness and flattening, one may compare asymmetry of various sets of histograms (9, 10).

Figs. $3a_1$ through $3f_1$ presented the histograms of the distribution of the numbers of various classes of actons for structures of the hind-leg of a horse (a_1), cow (b_1), pig (c_1), rabbit and dog (d_1), hen (e_1) and man (f_1). In Figs $3a_2$ to $3f_2$ are given the distributions of the frequency of various activities of actons. The structural schemes for their extremities at the mobility from $w = 8$ to $w = 30$ are given in Figs. $3a_3$ through $3f_3$. These histograms may also be analysed and compared for symmetry and degree of flattening.

2.3 Method of flattening and skewness evaluation of histograms of distributions of number of activities of actons

The degree of the flattening of the histogram may be easily expressed by coefficient γ_2, described by moments of 2nd and 4th order

(1)
$$\gamma_2 = \beta_2 - 3$$

where

$$\beta_2 = \frac{\mu_4}{\sigma^2} \quad ; \quad \bar{x} = \sum_{(i)} x_i p_i \quad \text{the mean value} \quad \sum_{(i)} p_i = 1$$

$$\mu_2 = \sigma^2 = \sum_{(i)} (x_i - \bar{x})^2 \quad \text{the variance}$$

$$\mu_4 = \sum_{(i)} (x_i - \bar{x})^4 \quad \text{the moment of 4th order}$$

In these calculations, p_i stands for the number of the joints, and x_i stands for the number of various activities.

Since the coefficient γ_2 is compared with the normal distribution, for which $\beta_2 = 3$, hence we get $\gamma_2 = 0$; for the uniform distribution $\gamma_2 = -1,8$. For positive values of γ_2 ($\beta_2 > 3$) one obtains a skewed distribution; for $\beta_2 < 3$, negative values of γ_2 are obtained i.e. the distribution is flattened. A very convenient measure for evaluating obliqueness is the coefficient β_1, described by the moment of 3rd order

$$\beta_1 = \frac{\mu_3}{\sigma^3} \tag{2}$$

where

μ_3 is the 3-rd moment.

For $\beta_1 = 0$ the set is symmetrical, for $\beta_1 > 0$ the set is skewed to the right side and for $\beta_1 < 0$ the set is skewed to the left side.

Applying this procedure to the histograms of the actions of the actons (Figs. 2 and 3) the relations $\gamma_2 = f(F)$ and $\beta_1 = f(F)$ have been found. Here, F - is the number of various activities, illustrated in Figs. 4a, b, c and d.

2.4 Relation between the number and the class of muscle actons

The results of the analysis made for 6 upper extremities of 5 mammals and a bird, concerning the relation between the number and the class of muscle actons, are illustrated in Fig. 5.

Fig. 5b shows a pictorial presentation of the analysis. The mean relations between these quantities is given in Fig. 5c.

The results of a similar analysis made for lower extremities are given in Figs. 6a, 6b and 6c. A comparison of Figs. 5c and 6c shows that both cases are characterized by similar relations.

2.5 Relationship between the number of degrees of freedom and the number of activities

The results of investigation of the relation between the number of degrees of freedom of individual extremities of 5 mammals and a bird versus the number of activities are presented in Fig. 7. It is evident from these results, that upper and lower extremities show a great regularity.

Only in case of the bird there is an exception (124 and 112 activities). Fig. 8a illustrates the relation: the number of degrees of freedom of the analysed structures of upper and lower extremities versus the number of fingers of the end of the link-chain. Fig. 8b presents the relation between the number of fingers of the end of the link-chain and the number of the activities, with the number of degrees of freedom fixed. Also in this case, the results show a definite regularity. The results of the analysis form a basis for a general procedure useful in structural and functional synthesis of extremity-like manipulators.

3. Introduction to the structural and functional synthesis of extremity-like manipulators and pedipulators

The initial data that are necessary for a synthesis include the anticipated functions i.e. manipulative or locomotive functions. These functions imply the fact that the manipulator (pedipulator) bears similarity to either upper or lower extremity. The initial data also include the number of various activities; the latter determines structural and functional complexity of the device.

3.1 Selection of the structure

After fixing the two above-mentioned factors, the number of degrees of freedom of the manipulator or pedipulator is read from Fig. 7. When this has been done, Fig. 8 helps to determine fingers of the hand of the manipulator. The following-stage is to describe the structure i.e. the number of links, kinematic pairs, their classes and their positions.

The structural analysis of extremities (Fig. $2b_3$ through g_3 and a_3 through f_3) shows, that the following classes of kinematic pairs appear: III, IV and V (rotational); the number of kinematic pairs $p_3 = 1$ (always) and its position corresponds to the shoulder or hip joint. Within the range of 22 degrees of freedom, only kinematic pairs of V-th class usually appear. Within the interval of degrees of freedom 23 to 30 also appear kinematic pairs of IV-th class, their number being less than 1/3 to 1/4 of the number of kinematic pairs of V-th class. These pairs usually appear in the upper and middle sections of extremities.

On the basis of this analysis a few approximate relations may be given; they are useful for determining the number of links, kinematic pairs and their classes in the manipulator or pedipulator.

For the case

$w \leqslant 22$ it can be assumed that

$p_3 = 1$, $p_4 = 0$ and $p_5 = n-1$

thus,

$$n = w-2, \text{ where}$$

$$n = n_1 + n_2$$

n_1 - the number of links of the connecting members

n_2 - the number of links of the end.

For upper extremity $\quad n_1 = 3 - 4$

For lower extremity $\quad n_1 = 2 - 3$

$$\text{and} \qquad n_2 = n - n_1$$

For $w \geqslant 23$ equation (2) leads to erroneous results. The analysis shows that in case of highly developed structures (for example: upper or lower human extremity) new formulae should be taken

$$p_3 = 1, \quad p_4 \leqslant \left(\frac{1}{3} \div \frac{1}{4}\right) p_5, \quad p_5 = \left(\frac{2}{3} \div \frac{3}{4}\right) n \qquad \text{and}$$

$$n = w-8, \quad \text{where}$$

$$n = n_1 + n_2 ; \quad n_1 = 4 \quad \text{and} \quad n_2 = n-4$$

3.2 Selection of drives for manipulator or a pedipulator

After deciding upon the structure of a manipulator or pedipulator, the designer should determine the number of drives, their functions and histogram of distribution in relation to the kinematic pairs (joints) that are serviced. For this purpose, the designer may utilize Figs. 9 or 10 to determine an adequate number of actons for a given number of degrees of freedom. In order to determine the distribution of muscles against activities the following procedure is suggested. On the basis of Fig. 11 the number of rectangles is determined. Then, utilizing curves plotted in Fig. 11, the designer should determine the degree of skewness of flatness of histogram. Now, taking advantage of formulae (1) or (2), the designer may determine the shares of various numbers of activities in servicing the joints. Utilizing the data included in Fig. 12 and 13 and in the set of equations (1) or (2), the designer may determine p_i, with $i = 2, 3$ and 4. If the histogram consists of more than 3 rectangles (for example 4 or 5) the calculated values of p_2, p_3 and p_4 should be supplemented with values of p_1 and p_5 obtained from additional premises of statical and controlling character.

Conclusions

The results of the above analysis may be utilized in a synthesis of extremity-like manipulators and pedipulators. Technical apparata with the structures and distribution of drives similar to living organisms should bring profit in the field of rational utilization of drives, and control and supply systems.

REFERENCES

[1]   K. FIDELUS Biomechaniczne parametry kończyn górnych czlowieka, Roczniki Naukowe AWF, TXV, PWN, 1971

[2]   K. FIDELUS , A. MORECKI - Niektóre wlasności biomechanizmów o zmiennej strukturze, Mechanika Teoretyczna i Stosowana 1, 9, 1971

[3]   A. MORECKI, K. KEDZIOR, K. NAZARCZUK, K. TEMPINSKI, K. FIDELUS , R. PASNIXZEK - Some Problems of Modelling and Measuring of Biological Systems, Proceedings of the Third World Congress for the Theory of Machines and Mechanisms, Kupari, Vol. E, 1971

[4]   A. MORECKI, J. EKIEL, K. FIDELUS - Bionika ruchu, PWN, 1971

[5]   A. MORECKI, K. FIDELUS - Niekotoryje swoistwa biomechanizmow pieremiennoj struktury, Maszinowiedienije 1, AN ZSRR, 1972

[6]   B. POPLEWSKI - Anatomia ssakow, T. I, II, III S.W.Czytelnik, 1948

[7]   J. GILL, Z. JACZEVSKI - Zarys anatomii i fiziologia zwierzat gospodarskich, PWRiL, 1968

[8]   FRANK W. CHAMBERLAIN - Atlas of Aviation Anatomy, Michigan, 1943

[9]   G.U. YULE, M.G. KONDALL - Watep do teorii statystyki, PWN 1966

[10]   O. LANGE, A. BANASINSKI - Teoria statystyki, PWE, 1968

Table 1.

Numerical presentation of activities of actons for joints of a foreleg of a horse.

Item	Name of the muscle acton	Shoulder Flex./Ext.	Shoulder Abd./Add.	Shoulder Pron./Sup.	elbow Flex./Ext.	wrist Flex./Ext.	intercarpal Flex./Ext.	M P Flex./Ext.	P i P Flex./Ext.	D i P Flex./Ext.	ki-class of a muscle acton	number of activ. of the act.
1	Latissimus dorsi	+	+	+							1	3
2	Teres maior	+	+								1	2
3	Supraspinatus c.mediale		+	+							1	2
4	Supraspinatus c.laterale		+	+							1	2
5	Infraspinatus	+		+							1	2
6	Teres minor	+		+							1	2
7	Subscapularis	+	+	+							1	3
8	Deltoideus	+									1	1
9	Brachiocephalicus	+									1	1
10	Pectoralis superficialis	+	+	+							1	3
11	Pectoralis profundus	+	+								1	2
12	Coracobrachialis	+	+								1	2
13	Biceps brachii	+			+						2	2
14	Tensor fasciae antebrachii	+			+						2	2
15	Triceps c. longum	+	+		+						2	3
16	Triceps c. mediale				+						1	1
17	Triceps c. laterale				+						1	1
18	Brachialis				+						1	1
19	Pronator teres				+						1	1
20	Flexor carpi radialis				+	+	+				3	3
21	Flexor digitorum superficialis				+	+	+	+	+		5	5
22	Flexor carpi ulnaris				+	+					2	2
23	Flexor digitorum profundus				+	+	+	+	+	+	6	6
24	Extensor carpi radialis				+	+	+				3	3
25	Extensor digitorum communis				+	+	+	+	+	+	6	6
26	Extensor carpi ulnaris				+	+	+				3	3
27	Extensor digit. accessoris					+	+	+	+	+	5	5
28	Abductor digiti I					+	+				2	2
29	Interosseus medius							+			1	1
	N. of muscles eng. in an activ.	5 8	2 7	4 3	7 7	6 3	5 3	3 2	2 2	1 2	17	7
	N. of activ. in a joint		9		14	9	8	5	4	3	5	11
	N. of activ. per one deg. freed.	13	9	7	14	9	8	5	4	3	3	7

72

0	0
2	2
2	2
58	72

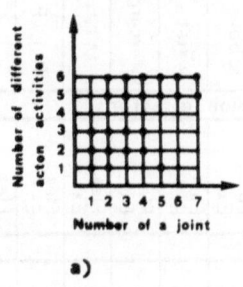

a)

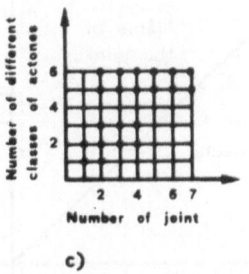

c)

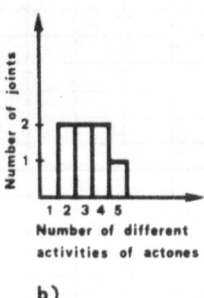

b)

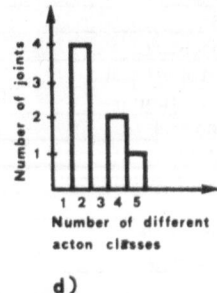

d)

fig. 1

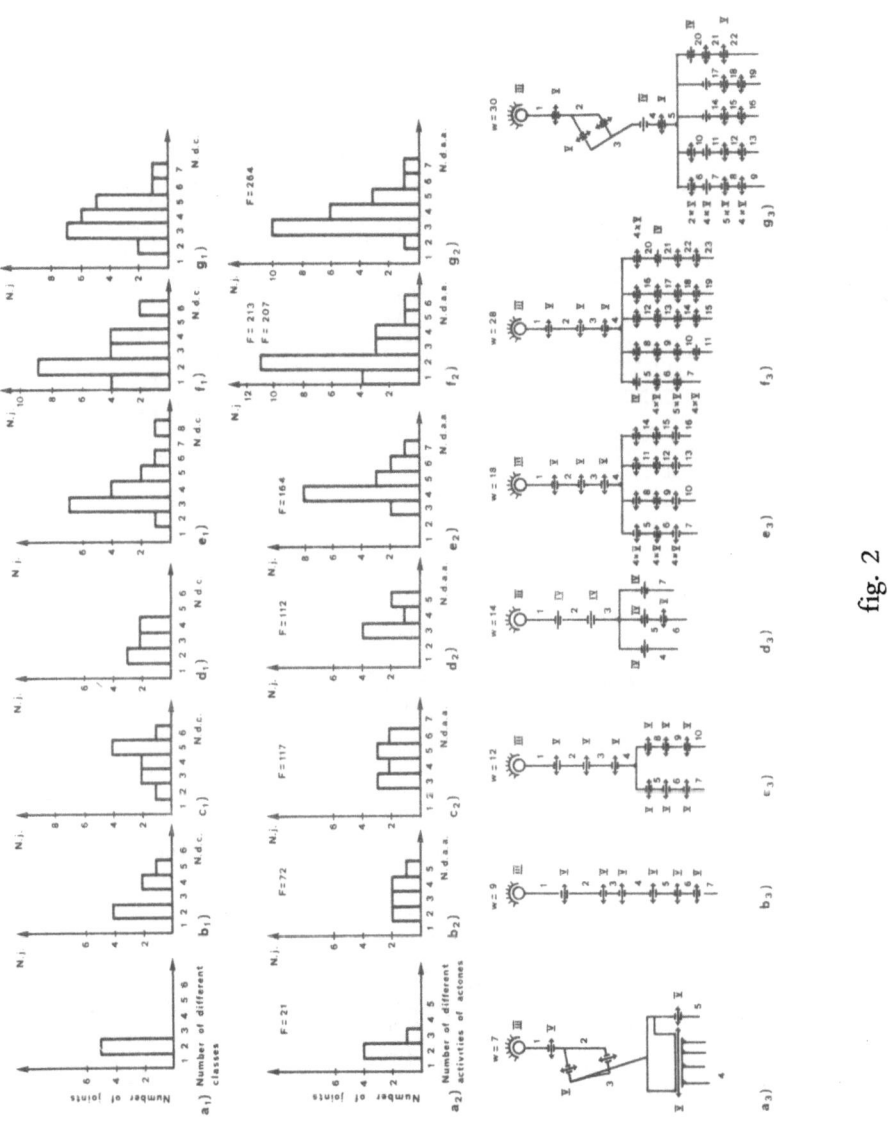

fig. 2

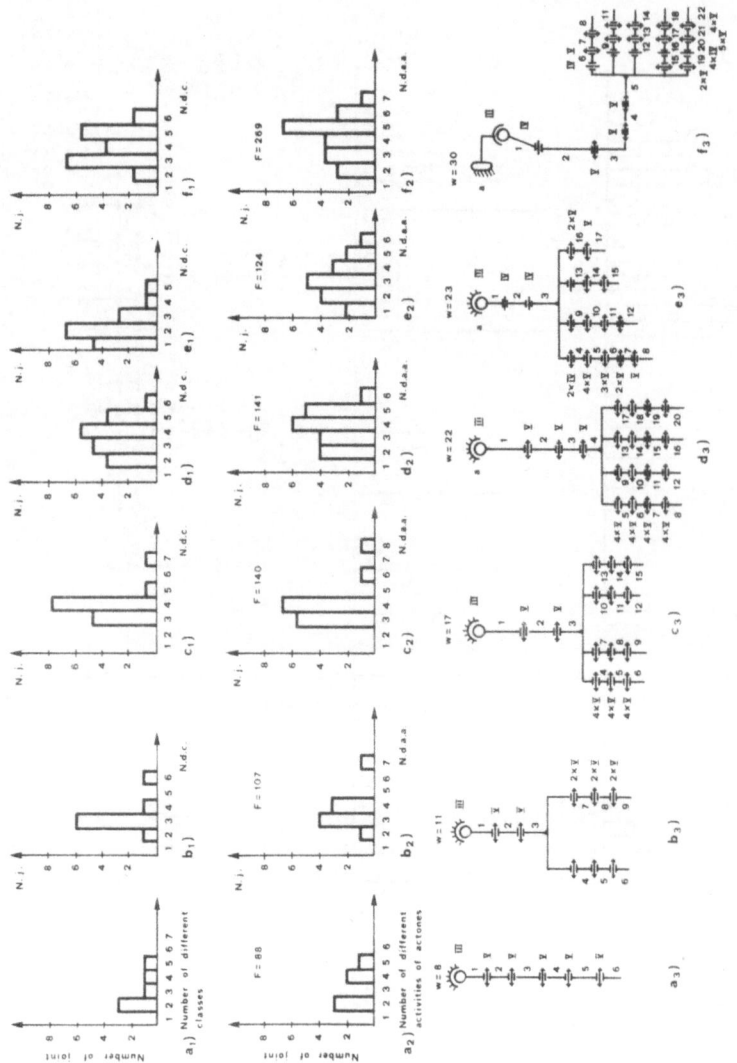

fig. 3

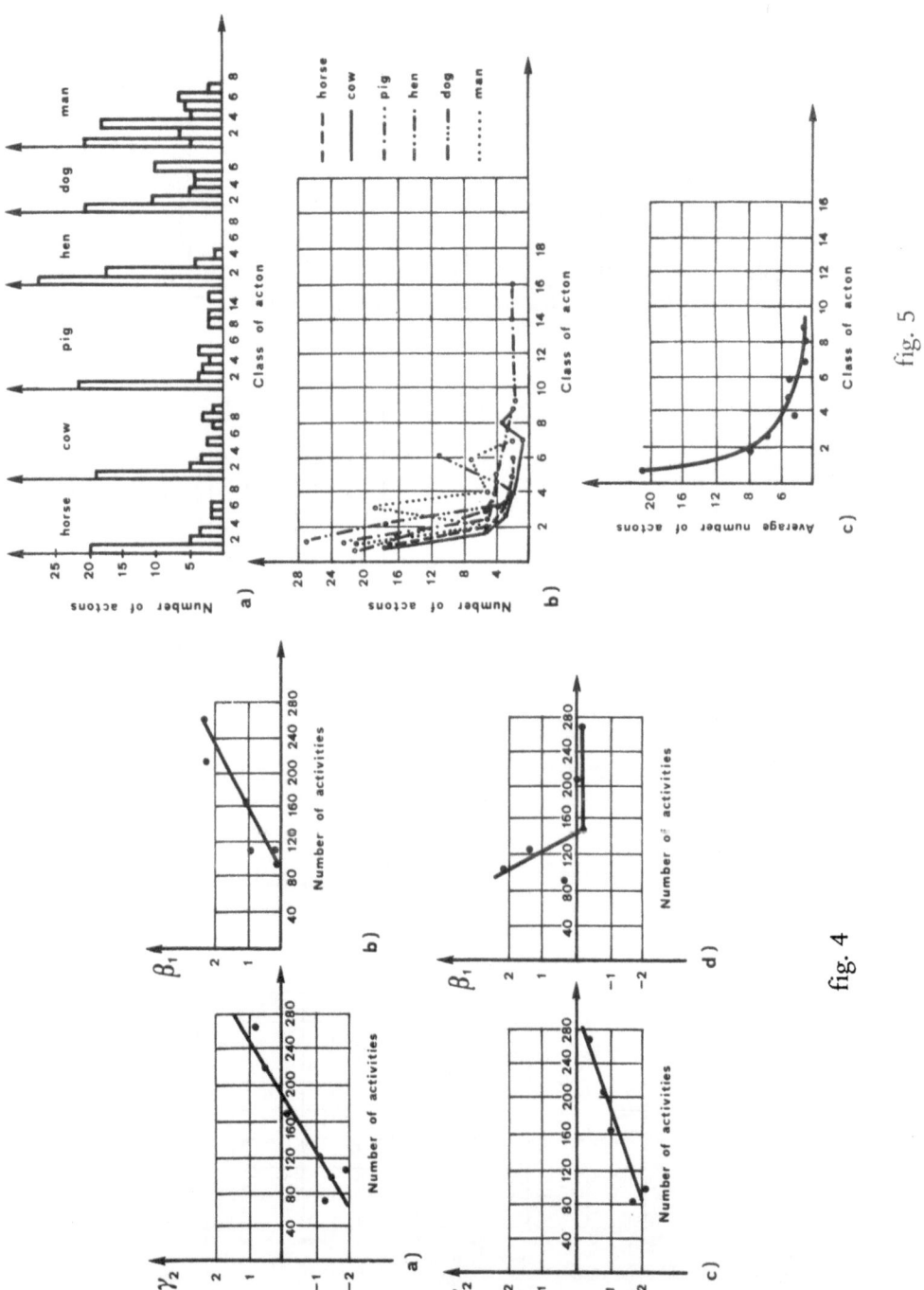

fig. 5

fig. 4

Number of actons
horse
cow
pig
dog
hen
man
Class of acton
a)
horse
cow
pig
hen
dog
man
Number of actons
Class of acton
b)
Average number of actons
Class of acton
c)

fig. 6

Upper extremity
Lower extremity
Number of degrees of freedom
Number of activities

fig. 7

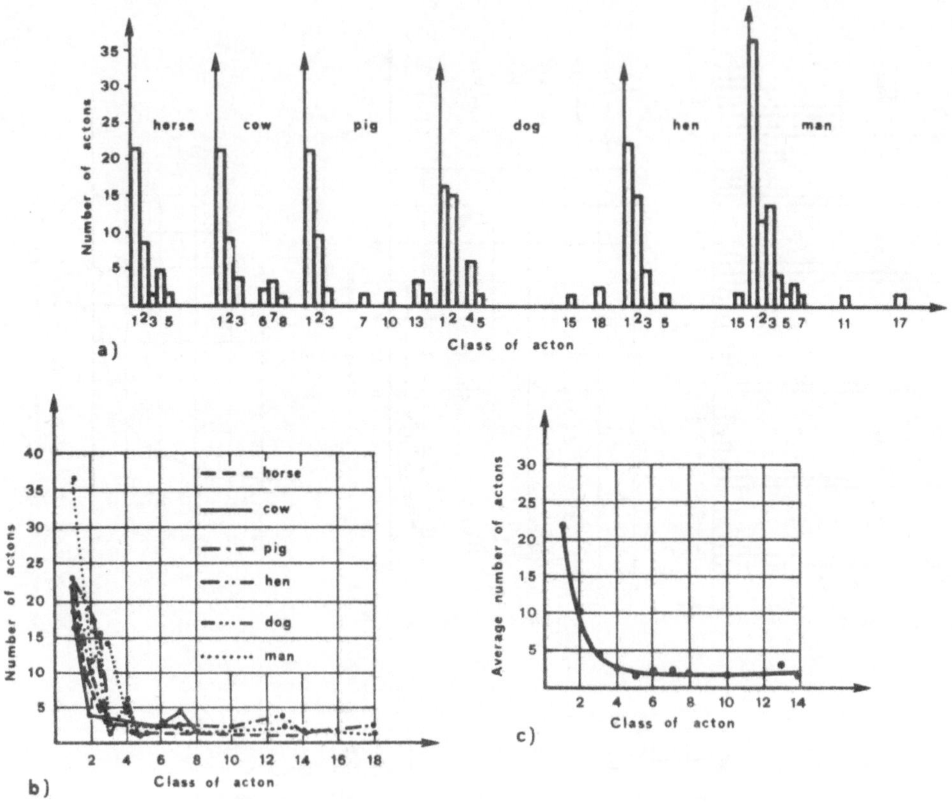

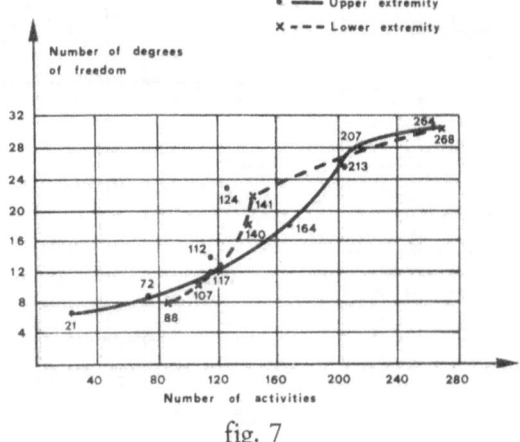

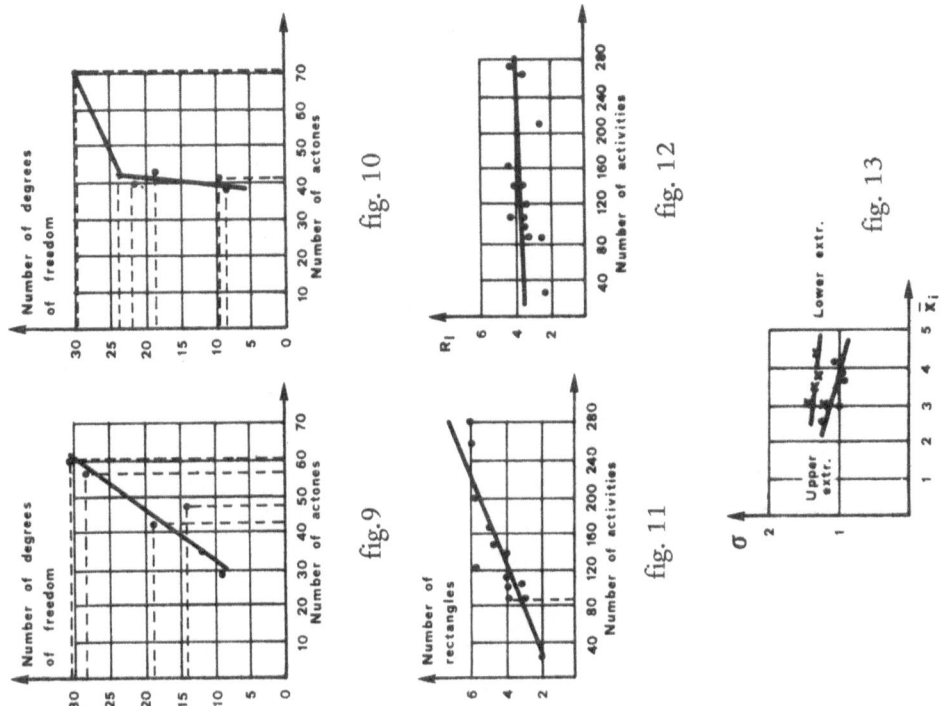

fig. 10

fig. 9

fig. 12

fig. 11

fig. 13

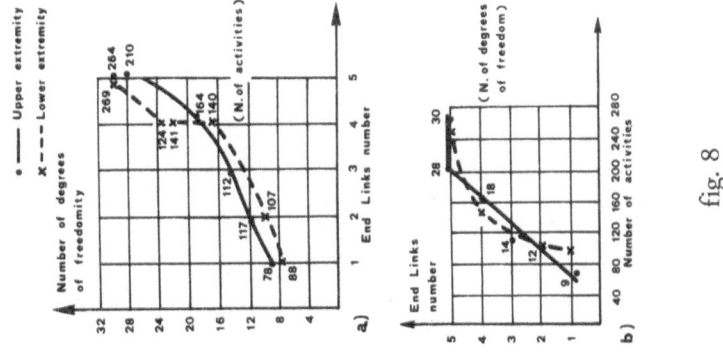

fig. 8

MEDICAL MANIPULATORS
(General considerations on the design criteria of manipulative technical aids for quadriplegics*)

Heinrich ROESLER, Dr. rer. nat.,
Orthopädische Klinik und Poliklinik
der Universität Heidelberg,
Heidelberg, W. – Germany.

Volkmar PAESLACK, Prof. Dr. med.,
Orthopädische Klinik und Poliklinik
der Universität Heidelberg,
Heidelberg, W. – Germany.

(**)

Summary

In a research program called medical manipulators the goal was set up to utilize the achievements of industrial manipulators and robots for the design of a functional aid for quadriplegics. An analysis of the basic properties of medical and industrial manipulators revealed so many differences that a completely new design was necessary.

The background of this new design was found by asking for the requirements of a number of quadriplegics. From their requirements a manipulation unit with 5 degrees of freedom was deduced which in combination with a rigorous rearrangement of the quadriplegics environment is able to perform the majority of the required tasks. General outlines for the design of control interfaces for two groups of quadriplegics with different control ability have been worked out.

(*) Sponsored by the Federal Ministery of Research and Technology

(**) All figures quoted in the text are at the end of the lecture.

1. Introduction

Despite of the urgent needs of a great number of severely handicapped people only a few attempts have been made to employ the achievements of sophisticated temporary mechanical and control technology to the development of manipulative technical aids. Many reasons might be adduced for this fact but the most important one is the difference in basic assumptions between a design for technical or industrial application and a design for the use as an aid for the handicapped.

In 1972 a research program was initiated by the German Federal Ministry of Research and Technology to evaluate the technological achievements in the field of industrial manipulators and robots for the design of medical manipulators. By the latter term a machine is described, which, completely or partially takes over the tasks of deficient human upper extremities of severely handicapped patients. A group consisting of medical people and engineers started working on this problem, and, after a time of discussing numerous proposals of approach, arrived at the conclusion that a simple transfer of design criteria of industrial manipulators was not feasible for the following reasons:

An industrial manipulator is designed for strictly defined tasks in appropriately established environments. Boundary conditions of its motions are task specific and design assumptions are represented by hard data, even if a high degree of versatility is demanded. The industrial manipulator is employed for only one or a few tasks with very many repetitions. Therefore it requires a flexible, easily changeable manipulation unit and a comparatively rigid task oriented programming of the control unit with one or very few facilities for an external interference into the automatically proceeding motion.

Tasks of a medical manipulator, on the contrary, are very difficult to define because of the vast repertory of motion patterns and the high adaptivity of the human model. The medical manipulator works in an a priori unarranged environment. Boundary conditions of its motions depend on the momentary task. A medical manipulator has to execute a large number of different actions with only very few repetitions. Its purpose as an aid for the handicapped requires a rigid and unchangeable design of the manipulation unit combined with a flexible control unit with many facilities for an external interference in the momentary motion. The control unit furtheron has to be adaptable to the individual abilities of the handicapped.

Because of these reasons in fact a completely new design concept was to be set up, the foundations of which had to be explored first. Some general considerations can be based on a rather simplified block diagram which represents the disabled human operator together with a medical manipulator in a random environment (fig. 1). Random environment in this context implies, that in the working region of the manipulator neither auxiliary equipment nor objects have been arranged such that in any kind manipulations are facilitated.

Internally generated intention and a certain momentary environmental situation are combined into a task. For the execution of this task information output is necessary which has to be transfered to the control interface and there transformed into control instructions for the manipulation unit. Then . the manipulation unit performs the expected movements, accomplishes the task and by this changes the environment, as is indicated by manipulation output in fig. 1. Feedback channels 1 and 2 display all sorts of information except optical one about the momentary state of the manipulation to the human operator, while feedback channel 3 shows direct optical supervision.

It is the peculiarity of the medical manipulator that the human operator is a disabled person. He is very well able to describe verbally a task to the last detail, but no language is available to transform this description into instructions for the machine. Disposing of only a few residual motor activities the number of information output channels of a disabled person is heavily reduced. Consequently the goal of the whole program must be to design a system, which, with a minimum of control instructions is able to perform a maximum of different tasks.

2. Conditions for the design of a manipulation unit.

The solution of a highly complex problem formulated in such a generality as above can only be found by setting up some boundary conditions. The first idea into this direction was that a medical manipulator which likewise meets the requirements of all types of disabilities would be an utopia at the present time. So as a special group of disabled people quadriplegics were chosen as possible users of medical manipulators. A further consideration was that a realization of the medical manipulator had to be carried through in reasonable time, which means a realization within several years. Therefore one had to aim at a combination as efficient as possible for the quadriplegic of already known developments in the field of the rehabilitation technology and for a first approach avoid time consuming new designs. In pursuing the development of the system after some experience with practical application more sophisticated solutions could be introduced. So the first approach was started with a close investigation of the situation of quadriplegics:

A quadriplegic suffers from a motor paralysis of all four extremities caused by a lesion of the spinal cord in its upper parts. In the worst case this means a complete loss of all motor capabilities with the exception of head and some shoulder movements. In less severe cases the patient disposes of several more motor functions but generally is not able to perform normal combined motions with his upper extremities. Locomotion is only possible by using a wheelchair. Since the disability of a quadriplegic is not progressing once the hospital treatment has come to an end, his residual motor functions can be judged in terms of degrees of freedom, the number of which specifies the number of independently available control sites. One degree of freedom is coordinated to the active mobility around one joint axis into two directions. The number of degrees of freedom can be connected with the hight of the lesion of the spinal cord.

By means of a questionnaire now data were collected about the special needs of quadriplegics, about their available control sites, their social and vocational life, their spare time activities, and their normal and vocational environment. The idea of a medical manipulator was described to them in a rather general way and they were summoned to set up a catalogue of tasks which they would expect of such a machine, and to utter their attitude towards this conception. They furtheron were informed that control would be rather difficult to learn and would require a large amount of personal engagement.

75 quadriplegics, preferably those with high lesions, were questioned. Their attitude towards the idea of a medical manipulator was correlated to their residual motor functions. If a quadriplegic disposed of 14 or more degrees of freedom, he tended to a negative attitude, if he disposed of less than this, he tended to a positive attitude. Apparently, as soon as the sufficient number of 14 or more degrees of freedom is still available, the variety of practicable motion patterns by use of residual motor functions becomes so large, that the quadriplegic thinks to be better off with his natural resources than with a medical manipulator.

From the various catalogues of tasks compiled by the quadriplegics now 68 actions could be extracted and a manipulation unit had to be selected which was able to perform these actions with a minimum of control instructions. The most simple type of manipulation unit under these conditions is a five degree of freedom version where the axis of rotation of the tool is permanently kept parallel to the horizontal (fig. 2). This type was originally developed by D. Simpson as an upper extremity prosthesis for amelic children and is well known as the Edinburgh arm.

Three degrees of freedom are needed for positioning in space, one for the final position of the tool, and the last for prehension. Only five independent control sites are required for the control of this basic manipulation unit.

The next question to answer was whether this basic manipulation unit was able to perform the 68 actions mentioned above. In a motion pattern analysis it was at first found out, that with the utilization of all the motor functions of human upper extremities 39 of these actions could be executed with one arm, 29 with both arms.

The examination which of the 68 single and double handed actions might as well be performed by the basic manipulation unit showed a surprisingly intensive interaction between the performance of the manipulator and the state of arrangement of its environment. Before any action could be analysed, some assumption or other was necessary about the environment because practically no action was feasible in a normal every day environment. The performance of a manipulator on the other hand is strictly correlated to its number of degrees of freedom. The state of order of a normal daily life and vocational environment is adapted to the high dexterity of human upper extremities, which together dispose of 60 degrees of freedom, 30 for each side. As was mentioned earlier we learned by our investigation that approximately 14 or 15 degress of freedom is the very lower limit a quadriplegic can get along with after a certain time of practice. But still this is only possible by using an assembly of special auxiliary equipment and by arranging his environment properly. If one transfers this finding to the manipulator, a further reduction to only five degrees of freedom will inevitably lead to a nearly complete inefficiency of the machine if not the environment is totally rearranged and adapted to its poor dexterity compared with human upper extremities. So the question whether the five degree of freedom basic manipulation unit is able to perform a certain task must be put just the other way round: Can the environment be arranged so that the manipulation unit is able to perform its task?

Under this aspect all 68 actions required by the quadriplegics have been resolved into sequences of single movements which could be performed by one human arm. Thereafter each of the single movements was tested separately whether the region of the environment where it was expected to take place could be rearranged to make the performance feasible. The analysis showed that for 61 out of the 68 actions this was possible. The last 7 actions were predominantly based on finger mobility — like knitting — and they were not taken into further consideration.

So the result of this investigation can be summarized to the statement that the basic manipulation unit will meet the requirements of the first approach to a medical manipulator for quadriplegics provided their environment will be rearranged to allow purposeful actions of the machine. This rearrangement includes auxiliary equipment for special actions. Fortunately, to a quadriplegic adaption of his environment to his individual disability is nothing totally new and therefore one may expect that he will tolerate the new boundary conditions, too, when being fitted with a medical manipulator.

3. Considerations on the design of control interfaces

From the very beginning of the work on medical manipulators everybody engaged in the problem was well aware that control was the more difficult part to cope with. All the difficulties arise because of the tremendous difference between the biological and technical information processing systems for the control movements. Whenever control of manipulators or other machinary by human operators is looked at, one has to discriminate two types of man machine systems. In the first type the human operator has to concentrate mainly on the control process and his primary task is the control of his technical equipment. A large amount of scientific work is concerned with the reactions of the human operator under these conditions [1]. Far less attention has been focussed on man machine systems where manipulation with the machine is the human operators primary task, where he is concentrated on the progress of the work while control proceeds more or less unconsciously. Human upper extremities work this way and their control is a completely unconscious process except in those situations where new motion patterns have to be learned. When writting on a piece of paper, for instance, man is occupied with what he writes and not with control of his hand and fingers drawing letters. The control interface of a medical manipulator should therefore be designed to include as many properties of its biological counterpart as is possible. On the other hand time consuming solutions were to avoid until more experience has been gained and so the application of computer assisted control interfaces has been postponed for the present time.

A survey in terms of degrees of freedom over the control ability of those quadriplegics who excepted functional advantages from a medical manipulator was gained by the evaluation of the questionnaire. Two groups had to be discriminated, one disposed of 4 degrees of freedom and less, the other one disposed of 11-14 degrees of freedom. In judging the control ability of a quadriplegic head movements have been excluded for reasons which will be explained later. The first group with high lesions and low control ability showed no great enthusiasm to undergo any troublesome learning process. Those with lower lesions and a correspondingly higher control ability on the contrary were very eager to augment their functional resources by using a medical manipulator even if learning to control the machine turned out to be a tiresome process.

Because one has to discriminate clearly between the two groups of quadriplegics with different control ability two versions of the interface are required. The interface for those quadriplegics who dispose of 11 degrees of freedom can be based on the control principle developed by D. Simpson during his work on prostheses for the complete arm [2]. This control principle is called extended physiological proprioception (EPP) and makes use of the fact that, provided with a position servo mechanism, the movement of a control point of the human body can be transformed to the movemement of a coordinated part of the artificial limb. Because of the consistent relationship between the position of the artificial limb and the position of the control point proprioception is processed by the central nervous system as if it was originated in the prosthesis, i.e. outside the biological system. Permanent contact between the control point and the control element of the interface during all control processes must be maintained, however. Under these conditions an unconscious pattern of control motions can be built up during the learning phase which corresponds to a transformed motion pattern of the artificial limb, provided the different control points of the human body are always used for one and the same function of the artificial limb [2]. Although it has not been proved yet a certain probability exists that this control principle still holds when applied to a quadriplegic and a medical manipulator which is not attached to the human body as a prosthesis but standing apart from it. One of the questions still open is whether the precision of the control will be sufficient for the medical manipulator , because small corrections of the final position can in case of the prosthesis be performed by body movements which is not possible with a manipulator.

If only 4 or less degrees of freedom are available control of the medical manipulator implies far more problems. Although head movements represent a relatively large source of information output for control purposes one cannot dare to utilize it permanently. For a quadriplegic head movements serve as non verbal communication and a solution of the control problem which involves an interference between manipulation and communication should be rejected. Head movements, therefore, will only be used for those controls, where a permanent connection from the human operator to the interface is not required, for instance emergency off or other switches which not directly create motions of the manipulator.

This additional condition, of course, still further reduces the choice of control points, so that one has to retreat on the very last possibilities of utilizing motor functions of the mouth and tongue, although this interferes with the cosmetic aspect. Various combinations and arrangements of control elements will be tested in the course of the time. One observation was made with all those interfaces hitherto tested: the coordination of a control point to the movement of a joint of the manipulation unit, as it was chosen for the Rancho arm aid and its tongue switch [3, 4], was found out to put nearly intolerable difficulties on the control process. Therefore it was decided that in cases where not sufficient control points were available and where one has to go back to sequential control, the control point has to be coordinated to a direction in space and not to a joint. The possibility to direct the terminal device by commands like forward – backward, up-down, left-right saves a considerable amount of operation time in all positioning tasks compared to the control of the single joint. This solution for the control interface will furtheron be better adapted to the basic manipulation unit.

4. Conclusion

At the momentary state of the investigation hardware can not yet be presented. The final construction of the manipulation unit and the control interfaces is under way and experience with the application of the system for the rehabilitation of the quadriplegics with high lesions will be gained in the near future.

The value of the findings hitherto collected in the course of the investigation can be seen in the augmentation of general knowledge on functional aids. The applicability of a functional aid does not only depend on the reliability, versatility, and dexterity of technology alone. It furtheron does not only depend on the properties of the man machine system in which the human operator might have gained enormous skill in controlling his device, but which, with the present knowledge of mechanical machines and of control on one hand, the heavily reduced information output of the disbled person on the other, cannot replace more than a very small part of the functional ability of the human arm and hands in a normal random environment. Even under optimal conditions the man machine system is condemned to partial futility if not the environment of the system has been adapted to the systems reduced functional versatility.

The statement of a disability is not an absolute one, it can only be stated in relation to the arrangement of the environment. This means that by changing a certain detail in the arrangement of the environment a functional disability can be provoked or, vice versa, made to disappear. Therefore a solution of the problem to replace upper extremities of a quadriplegic by a medical manipulator cannot be found in setting up a properly designed piece of hardware. A solution of the problem consists of hardware and a detailed program which includes a strategy for the rearrangement of the environment of the disabled with regard to his special deficiencies and an instruction how to practice new motion patterns with the manipulator in a rearranged environment by utilizing the residual motor functions of the handicapped.

REFERENCES

[1] BERNOTAT R.K., GAERTNER K.P., "Displays and controls„Swets and Zeitlinger N.V., Amsterdam, 1972.

[2] SIMPSON D.,"An externally powered prosthesis for the complete arm,,.The institution of mechanical engineers, Proceedings 1968-69, vol. 183, part 3J, pp. 11-18.

[3] JOHNSEN E.G., CORLISS W.R.,"Teleoperators and human augmentation„ NASA SP-5047, Washington, D.C., 1967.

[4] "Advancements in teleoperator systems, NASA SP-5081„ Washington, D.C., 1970.

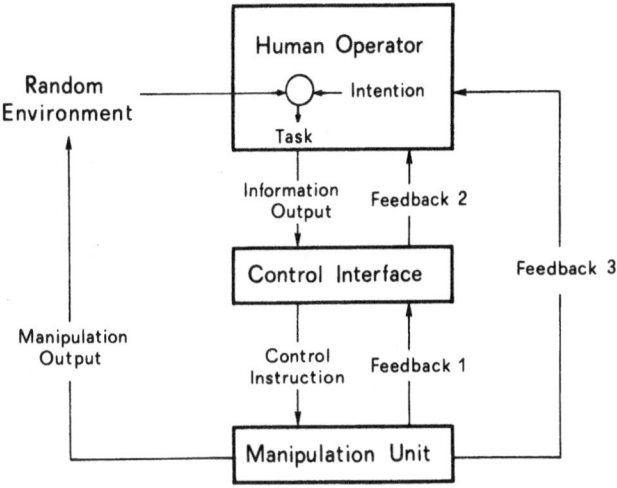

Fig. 1 Block diagram of a disabled human generator
with medical manipulator in a random environment.

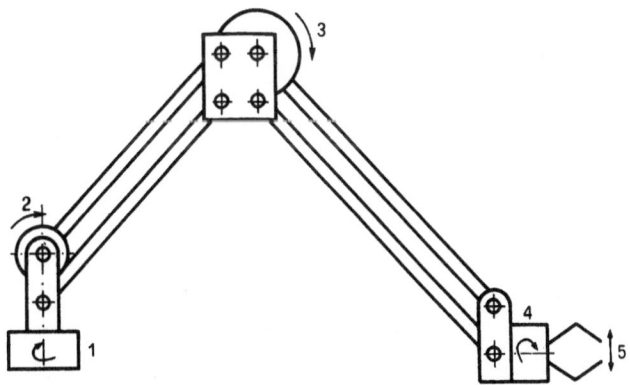

Manipulation unit with 5 degrees of freedom

Fig. 2 Basic manipulation unit with five degrees of
freedom for quadriplegics.

BIOMECHANICAL STUDY OF SERPENTINE LOCOMOTION

Yoji UMETANI, Associate Professor,
Tokyo Institute of Technology, Department of
Physical Engineering, Tokyo, Japan

Sigeo HIROSE, Graduate Student,
Tokyo Institute of Technology, Tokyo, Japan

(*)

Summary

The snake, in spite of its simple shape, can be considered to be a versatile robot having various functions. The authors are greatly intrigued by this fact, and believe that its biomechanical study will be instructive to design a new snake-like vehicle.

This paper deals with the kinematical analysis of the serpentine movement of snakes, zoological experiments for verifying the theory, and a mechanical model produced for designing snake-like vehicles for use as a robot.

(*) All figures quoted in the text are at the end of the lecture

1. Summary

This paper deals with the kinematical analysis of the serpentine movement of snakes, zoological experiments for verifying the results of the analysis, and a mechanical model produced for designing snake-like vehicles for use as a robot.

The body of a snake acts as a "foot" when it glides on the ground, as a "hand" when it coils around an object, and as a "trunk" when it moves from branch to branch on a tree. Thus the snake, in spite of its simple shape, can be considered to be a versatile robot having various functions. The authors are greatly intrigued by this fact and believe that a biomechanical study will suggest how to design a new animal-like versatile robot. In this paper, "serpentine movement" corresponding to the function of a "foot" is discussed.

The results obtained are as follows:

1) Fundamental kinematical relations for serpentine movement governing the straight-forward locomotion of snake are formulated.

2) Zoological experiments for verifying these relations with Elaphe Quadrivirgata are presented. They show good agreement with the theoretical analysis.

3) A peculiar habit of a snake's movement which the authors define as "sinus-lifting locomotion" (shown in the fig. 6), is an optimal mode for rapid movement.

4) The adaptive nature of a snake's undulatory motion with respect to ground friction can be explained from the kinematic relations.

5) A mechanical, active model made up of links, actuated by electrical servo-mechanisms has been successfully produced, and instructive knowledge for designing artificial serpentine vehicles can be obtained from it.

The authors think that this paper will be the first giving some biomechanical explanation of how snakes move and showing also the possibility of making snake-like versatile robots.

2. Kinematics of serpentine locomotion with zoological experiments

2.1. The gliding shape of a snake

The undulatory shape of a gliding snake must be quantified to analyze the kinematics of the locomotion. The authors found that two kinds of geometrical curve can represent the shape of a gliding snake; one is the clothoid curve and the other what we call the "serpenoid" curve. The formulas for these curves will be given here without any derivation.

Let the x-y coordinates be defined as shown in Fig. 5. Then one of the shape of a gliding snake can be expressed by smoothly connecting particular segments of the clothoid curve, which is formulated below with an auxiliary variable s, the elongated distance along the vertebrae of the snake.

$$\begin{bmatrix} x(s) \\ y(s) \end{bmatrix} = \ell \sqrt{\frac{\pi}{2\alpha}} \begin{bmatrix} \cos\alpha & -\sin\alpha \\ \sin\alpha & \cos\alpha \end{bmatrix} \begin{bmatrix} 1 & 0 \\ 0 & -1 \end{bmatrix} \begin{bmatrix} C\left(\frac{1}{\ell}\sqrt{\frac{2\alpha}{\pi}}\,s\right) \\ S\left(\frac{1}{\ell}\sqrt{\frac{2\alpha}{\pi}}\,s\right) \end{bmatrix} \tag{1}$$

where $C(*)$ and $S(*)$ are functions defined as the Fresnel's integral, and the other symbols are listed in Table 1.

Equation (1) is derived assuming that the contraction of the muscle increases or decreases monotonously and then the bending angle will vary proportionally along the vertebral axis s.

The other shape assumed is formulated as follows :

$$\begin{cases} x(s) = sJ_0(\alpha) + \dfrac{4\ell}{\pi} \sum_{m=1}^{\infty} \dfrac{(-1)^m}{2m} J_{2m}(\alpha) \sin\left(m\pi \dfrac{s}{\ell}\right) \\ y(s) = \dfrac{4\ell}{\pi} \sum_{m=1}^{\infty} (-1)^{m-1} \dfrac{J_{2m-1}(\alpha)}{2m-1} \sin\left(\dfrac{2m-1}{2}\pi \dfrac{s}{\ell}\right) \end{cases} \tag{2}$$

where $J_n(*)$ means the Bessel function of the first kind. This equation is obtained under the assumption that the muscle will contract repeatedly in such a way that the curvature of the vertebral line varies sinusoidally. This new type of Equation (2) seems not to have any name, and so the authors call it "serpenoid".

To verify the propriety of the shape of these curves, they were compared with those observed in living snakes (Fig. 1). These experiments showed that both the shapes given by Equations (1) and (2) are remarkable similar to that of a living snake. But, from the physiological point of view, the serpenoid curve is believed to be superior to the clothoid curve for expressing the mathematical shape of gliding snakes, because the muscle will naturally contract smoothly without any abrupt change.

2.2. The fundamental kinematic relations for serpentine locomotion

In analyzing the kinematics of snake locomotion, a chain model is assumed which is made up of active links connected by rotation joints as shown in Fig. 2.

Antagonistic muscles which cause the undulating bending motion of the whole trunk are also assumed to act as contracting actuators attached to both sides of a pair of adjacent links.

By the contraction force f_{mi} on one side of the antagonistic muscle, the force f_i ($= a/\delta s \; f_{mi}$) acts on joints J_{i-1}, J_i, J_{i+1}, as shown in Fig. 2. Thus, the resultant force acting on joint J_i is composed on the muscular contraction forces f_{mi-1}, f_{mi}, f_{mi+1}, while the effect of these forces on other link units does not need to be considered.

The resultant force on joint f_i is separated into two components, the tangential and the normal force. Referring to Fig. 2, the tangential force f_{ti} is calculated as

$$(3) \qquad f_{ti} = \frac{1}{2} \{ (f_i - f_{i+1}) + (f_{i-1} - f_i) \} \sin \left(\frac{\theta_i}{2} \right)$$

As the bending angle θ_i can be assumed to be small, $\sin(\theta_i/2) \simeq (\alpha/c) \cdot (x/\ell)$ will hold approximately. Therefore, Equation (3) can be reformulated into the following continuous equation,

$$(4) \qquad f_t(s) = - \frac{a\alpha}{\ell^2} \frac{dF_m(s)}{ds} s\delta s$$

where $F_m(s)$ shows the amount of muscular tensile force along the s-axis which must be determined from physiological considerations.

The total tangential force acting from O to P, a quarter part of an undulating cycle, can be derived by integrating equation (4) from O to P, to obtain;

$$(5) \qquad f_{t \, OP} = - \frac{a\alpha}{\ell^2} \int_0^\ell \frac{dF_m(s)}{ds} s \, ds$$

The normal force f_{ni} acting on joint J_i and the total normal force f_{nOP} acting from O to P can also be obtained in a similar way.

It is possible to determine the function $F_m(s)$, for it is based on the fact that i) muscular tension tends to zero at P, and ii) it increases monotonously from zero at P, to a maximum at O.

The authors propose this comparatively simplified function

$$(6) \qquad F_m(s) = f_{m0} \left\{ 1 - \left(\frac{s}{\ell} \right)^n \right\}$$

Though the parameter n, by which the characteristics of $F_m(s)$ is modified, is still unknown, it can be determined from physiological experiments. The kinematic equations for both the normal forces and the tangential forces are summarized in Table 2.

2.3. Zoological experiment

In order to verify these derived equations, zoological experiments were made using a common kind of snakes, Elaphe Quadriviragata, and the experimental results showed good agreement with the theory.

The muscular force was measured by EMG, which is believed to indicate equivalent values well correlated with the tensile strength of the muscular force. Besides the electrode for EMG, a newly developed normal force detector was attached to the snake as shown in Fig. 3. The photos of the gliding shape were simultaneously taken by a motor-driven camera to identify the position and the velocity of the snake.

Typical experimental oscillograms are shown in Fig. 4, and the synchronized photo corresponding to one instant (indicated by #) is shown in Fig. 5.

2.4. Further considerations

The results of the EMG and the normal force distribution show that the parameter n must be assumed to be in the range between 1 and 2. And it is also recognized that n will tend to unity when the snake moves as rapidly as possible. From Equations (5*) and (10*), we can derive the relation between the tangential and the normal forces.

$$\frac{f_t}{f_n} = \frac{2}{n+1} \alpha \tag{7}$$

In order to propel itself forward in serpentine movement, both the following conditions should be satisfied; $f_t \geq \mu_t W$ for producing tangential frictional slip, and $f_n \leq \mu_n W$ to prevent skidding in the normal direction. Thus equation (7) yields

$$\alpha \geq \left(\frac{n+1}{2}\right) \frac{\mu_t}{\mu_n} \equiv \alpha_0 \tag{8}$$

where, α_0 is defined as the lowest limit of the wriggling angle α. This relation (8) seems to represent one of the locomotion rules, that is, there exists a lower limit angle α_0 which the snake should maintain in order to move as fast as possible. Then, for a fixed ratio μ_t / μ_n, the parameter n should consequently be smaller to make α_0 small, and accordingly to make the motion more efficient. On the other hand, form muscular physiology we know that n is greater than 1, and so the parameter n should be kept at a value n = 1 + ϵ (where ϵ is small positive number).

The validity of this conclusion was also confirmed by the agreement between the observed propulsive force (110 [g]) and the theoretically calculated force (122[g]).

3. The adaptivity of the gliding shape

3.1. Explanation of the Sinus Lifting

An interesting habit of a snake when gliding on the ground as shown in Fig. 6, is what the authors have called the "sinus-lifting". This can be explained theoretically using some result, Eq. (8); the parameter n in the muscle distribution function should be taken to be $n = 1 + \epsilon$. Then the tangential force along the vertebral axis will be distributed according to the cusp-like curve shown schematically in Fig. 7. The authors have ascertained that such a distribution shows fairly good agreement with actual zoological experiments. Therefore, it seems to us that the snake concentrates its body weight around O, the portion that skids the most in order to obtain the most efficient gliding action, for the frictional force exerted on the surface to prevent a skid movement becomes greater as the compressing weight increases.

3.2. Adaptation of gliding shape to the surface

As shown by Equation (8), the wriggling angle α is equal to or larger than the critical angle α_0 ($\equiv ((n+1)/2) \mu_t / \mu_n$), depending on the surface, and the parameter n is fixed at $n = 1 + \epsilon$. Consequently, we may assume that the snake moves by adapting its gliding shape and adjusting the angle α to the nature of the surface.

To test the effect of varying the frictional coefficient ratio μ_t / μ_n of the surface, a gliding surface was constructed that could be inclined. The snake was made to glide up and down the slope, while cinema photos were taken from a position perpendicularly above the snake.

The equivalent frictional coefficient ratios μ_t^* / μ_n^* are calculated as follows.

Assuming a serpenoid curve

(9)
$$\frac{\mu_t^*}{\mu_n^*} = \frac{\mu_t \cos\theta + J_0(\alpha) \sin\theta}{\mu_n \cos\theta - \dfrac{4}{\pi} \sum_{m=1}^{\infty} \dfrac{J_{2m-1}(\alpha)}{2m-1} \sin\theta}$$

Assuming a clothoid spiral

$$\frac{\mu_t^*}{\mu_n^*} = \frac{\mu_t \cos\theta + \sqrt{\frac{\pi}{2\alpha}}\left(\cos\alpha C\left(\sqrt{\frac{2\alpha}{\pi}}\right) + \sin\alpha S\left(\sqrt{\frac{2\alpha}{\pi}}\right)\right)\sin\theta}{\mu_n \cos\theta - \sqrt{\frac{\pi}{2\alpha}}\left(\sin\alpha C\left(\sqrt{\frac{2\alpha}{\pi}}\right) - \cos\alpha S\left(\sqrt{\frac{2\alpha}{\pi}}\right)\right)\sin\theta} \tag{10}$$

The relations between the wriggling angle α and the frictional coefficient ratio μ_t / μ_n are summarized in Fig. 8. From this we can conclude that the snake adapts its movement to the surface by adjusting the wriggling angle α according to Equation (8).

4. Snake-like model vehicle and its control

In order to demonstrate the possibility of building a snake-like locomotion vehicle, a mechanical model has been successfully produced which is controlled like an automobile.

4.1. The mechanical model

The mechanical model, a snake-simulator vehicle, is shown in Fig. 9. The model has the structure of the snake toy Japanese children played with in ancient times. It is about two meters long, and consists of twenty actuated links mounted with servomechanisms. The whole model system is divided into two parts. One is the vehicle itself, and the other is a controller which is placed on a cart from which an electric cable bundle is attached to the vehicle to provide control signals and electric power. The model can glide on a flat surface just like a snake, using a control method which involves a sinusoidal reference input applied to the servomechanisms and a shift circuit for introducing a time delay between each servo unit signal.

4.2. Control of gliding shape and speed

The gliding shape which is determined by the wriggling angle α and the magnitude of the undulation ℓ, as well as the gliding speed V are controlled by adjusting the delay time Δt of the delaying circuit and varying the amplitude A and frequency f of a function generator used for the reference input.

When the length of a unit is 10 cm as here, the speed of locomotion V is expressed as,

$$V = \frac{10}{\Delta t} \tag{11}$$

The undulatory magnitude ℓ depends on the relations $f = 1/4c. \, \delta t$ and $\ell = c \, \delta s$, as shown by:

$$(12) \qquad \qquad \ell = \frac{2.5}{f \cdot \Delta t}$$

The twisting angle α is, omitting the derivation, given as

$$(13) \qquad \qquad \alpha = 1.79 \, \frac{A}{f \cdot \Delta t}$$

These derived relations were verified in the experiment described in the following section.

4.3. Manoeuvring experiment

The manoeuvring experiment was performed on a level surface. The auxiliary cart carries the controlling unit and power sources and follows behind the self-propelling model. The mechanical model propelled itself by serpentine locomotion.

One set of the variables used in the experiment was : $\Delta t = 0.6$sec, $f = 0.12$Hz, $A = 1.5$V. The corresponding observed gliding shape variables $V = 17$cm/sec., $\ell = 35$cm, $\alpha = 40°$. These agree fairly well with those derived from Equations (11) (12) (13).

The authors were also able to control the model manually so as to make it move along a preassigned course.

REFERENCES

There are a few works on how snakes move. In [1] and [2] the anatomy and the mathematics, respectively, are dealt with and in [3] there is collected a wide range of zoological investigations of snake locomotion.

[1]   MOSAUER W.,"Locomotion of snakes and its anatomical basis,, Ph.D. Thesis, University of Michigan, 1931, pp. 117-147.

[2]   RASHEVSKY N.,"Mathematical Biophysics,, Vol. II, Dover Publ., Inc., N.Y., 1962, pp 256-261.

[3]   GRAY J.,"Animal locomotion,, 1968, Norton, pp. 166-193.

The authors' contributions relevant to this paper are to be found in following two:

[4]   UMETANI Y.,"Kinematics and control mechanisms of serpentine movement — Part II. Analysis and experiment by a model,, BIOMECHANISM, University of Tokyo Press, Tokyo, 1972, pp. 243-251.

[5]   UMETANI Y., HIROSE S.,"Biomechanical study on serpentine locomotion — Mechanical analysis and zoological experiment for the stationary straightforward movement,, —, Trans. Soc. Instrument & Control Engrs., 8, 6, 1972, pp. 724-731.

Fig. 1. Comparison of theoretical curves with the observed shape of gliding snake (shaded portion indicates standard deviation around mean shape).

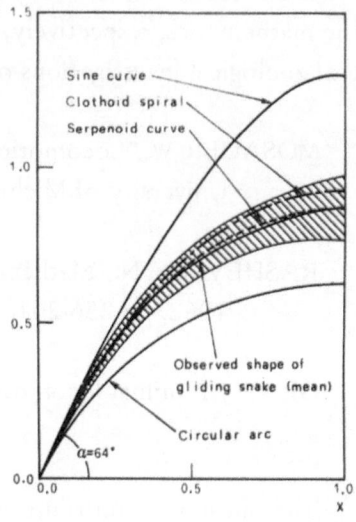

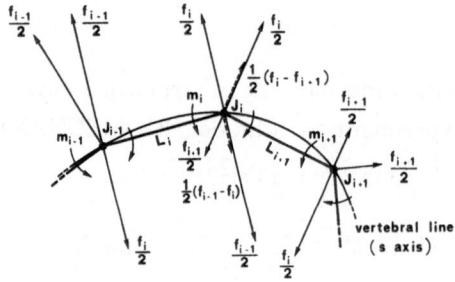

Fig. 2 Forces from muscular contraction acting on a joint J_i.

Fig. 3. EMG electrodes and normal force detector mounted on the snake (Elaphe quadrivirgata).

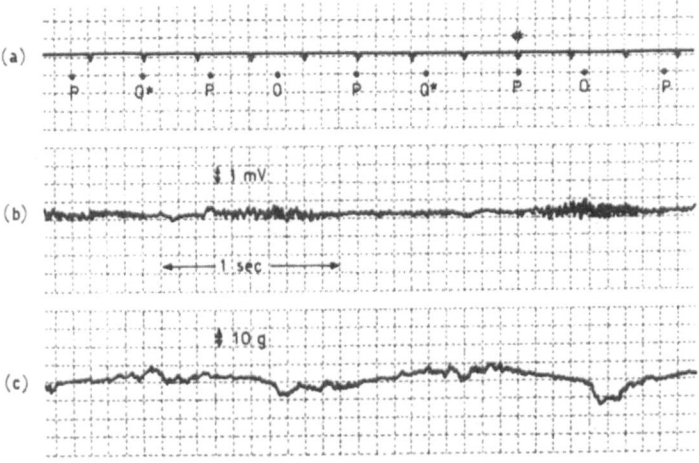

(a) photo synchronizing signal (Location of O,P)   (b) EMG   (c) normal force

Fig. 4. An example of experimental data (Time of photo in
fig. 5 is indicated by #)

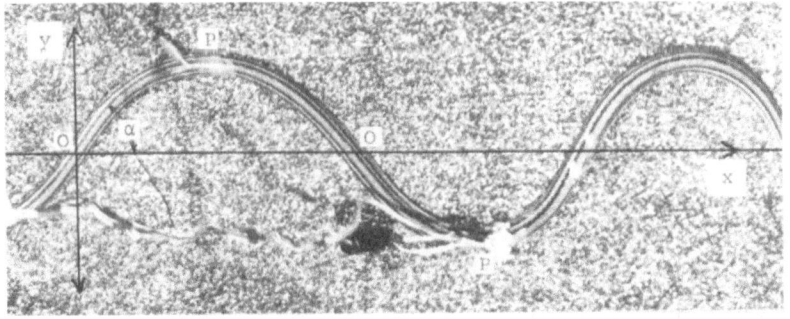

Fig. 5. One of the series of photographs (Position O, P,
angle α, x-y coordinates are illustrated)

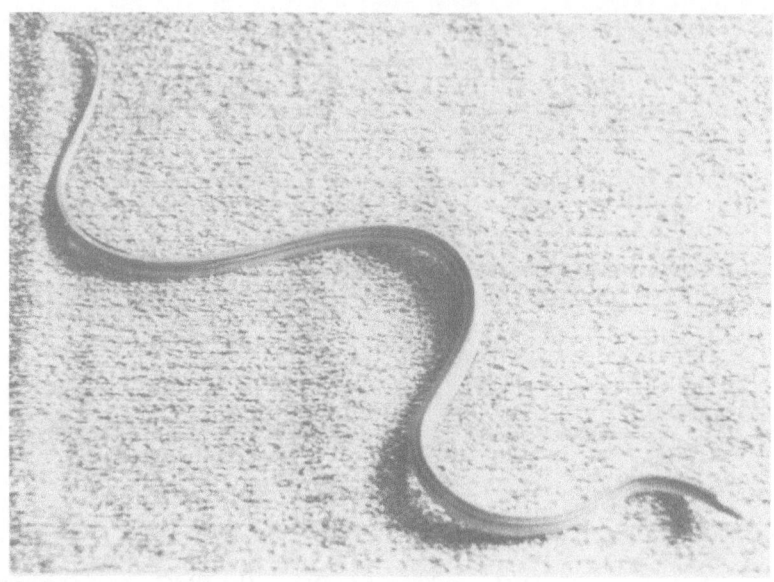

Fig. 6. Rapid serpentine movement of the snake (Elaphe qua-
         drivirgata)(note that the snake's body is raised from the
         surface of curvature peaks).

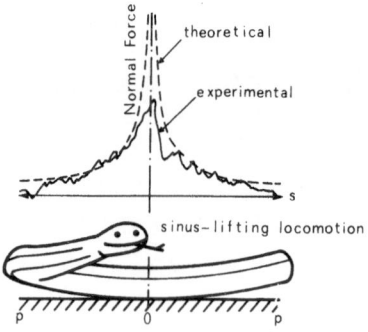

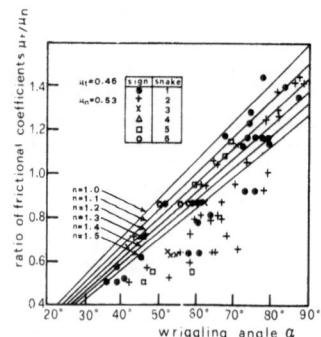

Fig. 7. Normal force distribution in
         sinus-lifting locomotion.

Fig. 8. Relation between μ_t / μ_n
         and α.

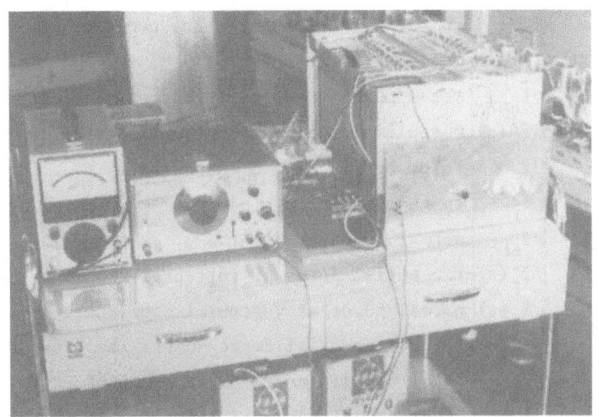

Commander (Controller and power sources)

Snake-like vehicle

Fig. 9. Mechanical snake-like model

TABLE 1 Notation

a	mean distance betwen joint and muscle
c	number of vertebrae in the range of $\overline{OP}$
f_{mi}	muscular force on joint J_i
f_{mOP}	total muscular force in the range of $\overline{OP}$
$F_m(s)$	muscular force distribution function
f_i	force on J_{i-1}, J_i, J_{i+1} induced by f_{mi}
f_{ti}	tangential force at joint J_i
f_{ni}	normal force at joint J_i
$f_t(s)$	tangential force at vertebral coordinate s
$f_n(s)$	normal force at vertebral coordinate s
f_{tOP}	total tangential force in the range of $\overline{OP}$
f_{nOP}	total normal force in the range of $\overline{OP}$
f_t	craniocaudal tangential force
f_n	craniocaudal normal force
L_i	the i-th link from point O
J_i	the i-th joint from point O
ℓ	distance along vertebral axis from O to P
O,P	positions shown in Fig. 5
s	vertebral length from point O
α	wriggling angle shown in Fig.5
δs	unit length of vertebra
$\delta\theta$	increment of bending angle
θ_i	bending angle betwen L_i and L_{i+1}

TABLE 2 Summary of derived equations

Force / Equation	$f_t(s)$ Tangential Force	f_{tOP} Total Tangential Force	$f_n(s)$ Normal Force	f_{nOP} Total Normal Force
Basic Equation	$-\dfrac{a\alpha}{\ell^2}\dfrac{dFm(s)}{ds}s\delta s$ (1*)	$-\dfrac{a\alpha}{\ell^2}\displaystyle\int_0^{\ell}\dfrac{dFm(s)}{ds}s\delta s$ (3*)	$-\dfrac{a}{2}\dfrac{d^2Fm(s)}{ds^2}\delta s$ (6*)	$-\dfrac{a}{2}\dfrac{dFm(\ell)}{ds}$ (8*)
Equation characterized by muscular distribution of Eq.(6)	$n\dfrac{a\alpha}{\ell^2}f_{mo}\left(\dfrac{s}{\ell}\right)^n\delta s$ (2*)	$\dfrac{n}{n+1}\dfrac{a\alpha}{\ell}f_{mOP}$ (4*) $\dfrac{a\alpha}{\ell}\dfrac{f_{mOP}}{c}$ (5*)	$\dfrac{n(n-1)}{2}\dfrac{a}{\ell^2}f_{mo}\left(\dfrac{s}{\ell}\right)^{n-2}\delta s$ (7*)	$\dfrac{n}{2}\dfrac{a}{\ell}f_{mo}$ (9*) $\dfrac{n+1}{2}\dfrac{a}{\ell}\dfrac{f_{mOP}}{c}$ (10*)

KINEMATIC FEATURES OF HUMAN ARM
IN "OPERATOR–MANIPULATOR" SYSTEM

I.B. VINOGRADOV, Junior Researcher,
Institute for the Study of Machines,
Moscow, USSR

(*)

РЕЗЮМЕ

    Дается поняте о методе кинематического анализа манипуляторов - методе объемом. Описывается эксперимент по определению ограничений подвижности в плечевом и запястном суставах руки человека. Приводятся методики исследования руки оператора и системы "оператор-манипулятор" методом объемов и результаты расчетов.

(*) All figures quoted in the text are at the end of the lecture

The conventional methods of kinematic analysis of mechanisms make it possible to study the kinematics of a manipulator by giving the laws of motion for all its major components. The difficulties caused by a great number of mobility degrees of the system are of a purely technical nature. However, such an approach based on the study of various trajectories of motion does not give any idea about general kinematic properties of the manipulator. Due to this reason it seems to be reasonable to use, along with such a trajectory method of kinematic analysis, another method — the method of volumes which enables one to obtain general integral estimates of manipulator kinematic properties, which characterize its peculiar qualities in the mastering of a set of trajectories.

The manipulator operates in a working space defined by constructional limitations on the lengths of manipulator members and on motions in its joints. The tong of the manipulator can reach any point of this space but the functional possibilities of the mechanism turn out to be different in different points. In other words, each point of the working space of the given manipulator possesses certain features characterizing possibility of fulfilment of various working operations. A study of the properties of the working space of the manipulator does not give exhaustive information about every specific working operator permited, however, it serves to estimate important kinematic characteristics of the manipulator as a whole.

During the operation of the manipulator it is necessary to have a possibility of orienting and moving the tong in different ways with respect to the object of manipulation. However, the structural and constructional limitations do not permit bringing the tong to every point of the working space at every angle. For each point of the space it is possible to determine a certain solid angle ψ, within which the tong can be led up to this point. Let us call the angle ψ the space maintenance or service angle. Let us call the relationship $\psi/4\pi = \Theta$ the coefficient of sercice at the given point.

Thus, it is possible to determine a certain scalar field over the working space, that is to put in correspondence to each point of the space its service coefficient. The value of this coefficient may vary from 0 for the points on the boundary of the working space to 1 for the points of the so called zone of 100 per cent or complete service. It seems to be correct to characterize the quality of the manipulator as a whole by the average value of the service coefficient Θ in the working space V.

$$\overline{\Theta} = \frac{1}{V} \int_V \Theta \, dv$$

which can be called the complete coefficient of maintenance or the "service" of the manipulator.

The study of slave arms of manipulators by the method of volumes is made in [1]. The same method can be used for studies of master arms of the master-slave manipulator. However, in this case it is necessary to consider the "operator-manipulator" system in which not only the features of the mechanical part of the system are taken into account but also the kinematic features of the human arm, which is an open kinematic chain with a great number of degrees of freedom. In the development of the methods for estimation of the service characteristics and in experimental research a simplified model was used (Fig. 1) - mechanical analog of the arm, in which the shoulder and wirst joints are represented by combinations of three mutually perpendicular turning pairs of the V class 2 with the axes crossing in one point, and the elbow joint – by the turning pair of the V class. Near the axes of hinges in Fig. 1 there are given designations of angles which correct the adjacent members, and the thin lines designate the original position of the kinematic chain at which all the angles are equal to zero. The angles φ_1, φ_2, φ_3 determine position of the shoulder with respect to body, the angle φ_4 -the position of the forearm with respect to shoulder, and the angles φ_5, φ_6, φ_7 - the position of wrist with respect to the forearm. Mobilities in breastbone-clavicular and acromioclavicular joints refer to the shoulder joint and are not taken into account separately. Thus, according to our approximate model, the configuration of arm is completely determined by the values of the generalized coordinates φ_i, where i $= 1, 2, \ldots, 7$.

Let us assume that the wrist fulfils only one function – the operator tightly envelopes the handle of the master unit of the manipulator in such a way that the wrist of the operator and the manipulator handle are as a single member.

In order to identify our rough model, at least in the first approximation, with the operator's arm it is necessary to attribute to the turning pairs the mobility limitations $\varphi_{i\min}^*$ and $\varphi_{i\,max}^*$ peculiar to the joints of the human arm. These limitations occur due to a number of reasons, which can be conditionally divided into "internal" connected with anatomical peculiarities of an individual joint and into "external" which are determined by specific features of arrangement of the working space (dimensions of the operator's body, the presence of stationary technological equipment, etc.).

Information cited in the literature [3, 4] on the mobility of arm joints turns out to be insufficient for studies of its mobility, since it characterizes maximal (amplitude) values of angles measured only for convenient positions of arms in horizontal, front and sagittal planes. In the multitude of configurations formed by the arms members these amplitude values rather often are inaccessible due to the presence of Interconnection between limiting positions over degrees of freedom of the shoulder or radiocarpal joints. The above mentioned features essentially influence the manipulative properties of the arm. Therefore prior to analytical estimation we had to determine experimentally the pattern of distribution of limitations of mobility $\varphi_1^* = f(\varphi_2, \varphi_3)$ in shoulder and $\varphi_7^* = f(\varphi_5, \varphi_6)$ in radiocarpal joints as functions of angles of mutual arrangement of members which form these joints.

For this purpose a special device has been developed realizing the scheme of the kinematic chain adopted in the model. The device is put on the arm of the operator in such a way that axes of all the hinges pass through the centers of rotation of the corresponding joints. Mobility in any of the turning pairs (every one of them is furnish with a limb for recording the values of angles) can be excluded.

The method for determining the limitations φ_i^* of mobility, for instance of the angle φ_1^*, reduces fixing the positions of the two other kinematic pairs (φ_2, φ_3) of the investigated joint. Then the ultimately accessible postions of the arm in moving in both directions (in our example $\varphi_{1\,min}^*$ and $\varphi_{1\,max}^*$) are registered with the aid of the measuring limb. A number of records of such limiting positions in successive consideration of fixed positions of two kinematic pairs permits the plotting of the unknown relationships of φ_i^*. Fig. 2 represents the obtained relationships $\varphi_1^* = f(\varphi_2, \varphi_3)$. It is seen that mobility limitations in the shoulder joints depend essentially on shoulder position defined by the angle φ_3. In the radiocarpal joint similar functional peculiarities are considerably less manifested. Fig. 3. represents experimentally determined values of the function $\varphi_7^* = f(\varphi_5, \varphi_6)$. The functions $\varphi_1 = f(\varphi_2, \varphi_3)$ and $\varphi_7^* = f(\varphi_5, \varphi_6)$ are approximated by Chebyshev orthogonal polynomials.

The calculation of the service coefficient is fulfilled only under the condition that the manipulator handle is connected with the given point of space through the spherical hinge. In this case shoulder, forearm and "wrist-handle" member form the closed kinematic chain 0Λ3C (Fig. 1) which has four degrees of mobility [2]. If the motions are fixed by the angles φ_3 and γ (the latter characterizes rotation around the handle axis) then we obtain a space mechanism whose driving member is the handle. Having given the latter the laws of motion over two degrees of freedom of the spherical hinge C, let us make the remaining members of the mechanism accomplish certain motions in the space, which are characterized by the generalized coordinates φ_i. The latter are calculated as follows. The coordinates of the wrist 3 and of the elbow Λ in the coordinate system X Y Z, and the angles φ_1, φ_2, φ_4 are calculated. Then the axes X Y Z are transferred to point 3 and with the aid of matrices of rotation are turned successively through the angles φ_1, φ_2, φ_3, φ_4. In a new system of coordinates X Y Z the coordinates of the points C, K and the angles φ_5, φ_6, φ_7 are searched.

For estimating the arm service coefficient let us make the handle perform two such motions that its axis CK would discretely move through uniformly distributed points of the sphere forming a bundle of straight lines in the center C. For each position of handle it is necessary to calculate φ_i and compare them with limitations over mobility in the joints $\varphi^*_{i\,min}$ and $\varphi^*_{i\,max}$. If the inequality is fulfilled

$$\varphi^*_{i\,min} \leqslant \varphi_i \leqslant \varphi^*_{i\,max} \tag{1}$$

then the position of the handle is considered to be real for the arm of the operator. If, however, inequality (1) is not fulfilled then the fixed values of the angles φ_3 and γ should be changed and all the calculations repeated. The position of the handle can be considered unreal only in the case when inequality (1) is not satisfied for all possible values of the angles φ_3 and γ. The bundle of real positions of the handle traces the angle of service in the given point. The service coefficient can be found from the ratio between the number h of real positions of the handle, and the total number N of discrete positions of its axis in the given point

$$\Theta = \frac{h}{N} \tag{2}$$

According to the developed methods calculations were made and the pattern of service coefficient distribution in the area of accessibility of the operator's hand was revealed. Fig. 4 shows approximate arrangement of zones A and B with the highest service coefficient for the right hand of one of the operators.

The studies of kinematics with the use of service characteristics can be also extend to the "operator-manipulator" system. Let us combine for this the slave and master arms of the manipulator (in the scale of the master arm) and connect to the master handle a kinematic chain simulating the operator's arm (Fig. 5), manipulator tong ℓ_3 being considered the driving member of the two kinematic chains.

Applying similar methods let us set up for the member ℓ_3 discrete motions over the sphere around the point C in such a way as to form in it a bundle of straight lines uniformly distributed over the sphere surface. For each position of the tong let us find the determine configuration of generalized coordinates of the manipulator ψ_j (where $j = 1, 2, \ldots, 5$) and of the operator's arm φ_i (where $i = 1, 2, \ldots, 7$). The position of the tong is real if

$$\psi^*_{j \, min} \leqslant \psi_j \leqslant \psi^*_{j \, max} \, , \quad \varphi^*_{i \, min} \leqslant \varphi_i \leqslant \varphi^*_{i \, max}$$

where $\psi^*_{j \, min}, \psi^*_{j \, max}$ are the limiting values of the generalized coordinates for the manipulator.

The coefficient of service in the point can be estimated by the number of real positions of the tong from relationship (2). Let us calculate Θ in the points at regular intervals over all the working volume of the manipulator. Then determine the arithmetic mean of the value of the service coefficient. This value we shall call the service of the "operator-manipulator" system. The given algorithm is programmed for computation on a digital computer. By varying the parameters of the master arm of the manipulator and estimating every variant of the "operator-manipulator" system by the value of the service it is possible to choose the most advantageous design of the manipulator.

ACKNOWLEDGEMENT

The author expresses his thanks to Professor A.E. Kobrinskii for consultations and support.

REFERENCES

[1] VINOGRADOV I.B., KOBRINSKII A.E., STEPANENKO Yu.A., TYVES L.I., "Some Problems of the Theory of Manipulators„ Proceedings of the II International Congress on the Theory of Machines and Mechanisms, 2, Warszawa, 1969.

[2] ARTOBOLEVSKII I.I.,"The Theory of Mechanisms„ Nauka, Moscow, 1967. (in Russian).

[3] WOODSON W.E., CONOVER D.W.,"Human Engineering Guide for Equipment Designers„ Univ. of California Press Berkeley, Los Angeles, 1966.

[4] "Human Engineering Guide to Equipment Design„ Sponsored by Joint Army-Navy-Air Force Steering Committee.

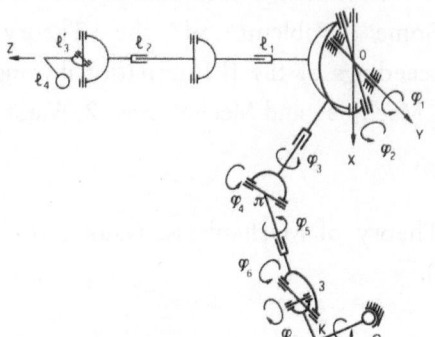

Fig. 1

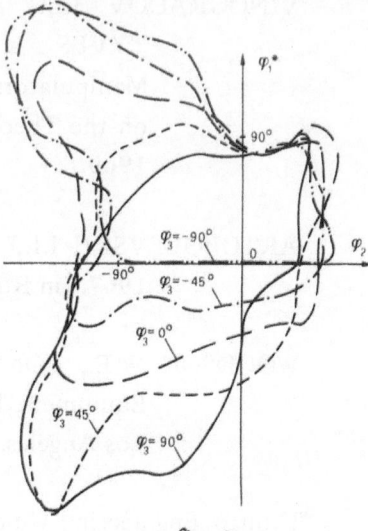

fig. 2

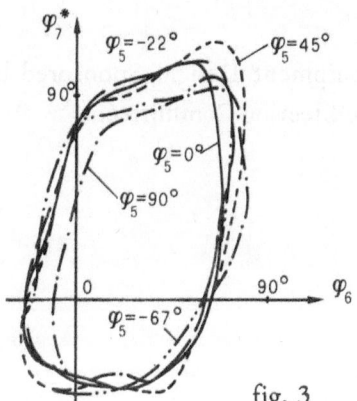

fig. 3

fig. 4

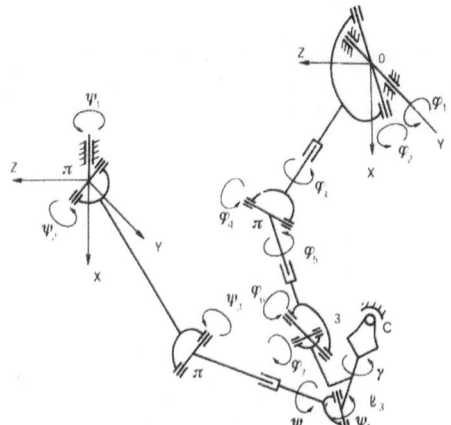

fig. 5

4. MAN–MACHINE SYSTEMS

THE APPLICATION OF SENSORY INFORMATION AND MULTIFUNCTION LEARNING TO AUTONOMOUS MANIPULATOR CONTROL

Amos FREEDY, Fred HULL, Gershon WELTMAN, and John LYMAN

Perceptronics, Inc. and
University of California, Los Angeles

(*)

Summary

This paper describes a computer system which can learn to control a remote manipulator, including the ability to search for goals and subtasks and to carry out fine handling motions. The system is an expanded version of the Autonomous Control Subsystem described in an earlier publication. Like the earlier concept, the new system begins by observing the control actions of a human operator, and subsequently takes over control responsibility. Autonomous control functions are provided by a set of interconnected learning networks which relate the sensory experiences of the machine to the generation of arm and hand motions. For remote applications the computer can be pretrained to operate autonomously in a simulated environment and will adapt to unexpected changes in the actual working environment. The paper presents the mathematical basis of system design, and concludes with a discussion of advantages.

(*) All figures quoted in the text are at the end of the lecture.

Introduction

An earlier publication introduced the concept of shared manipulator control between the human operator and a computer based learning system (Freedy et al. 1971). The concept features a trainable control network termed the Autonomous Control Subsystem (ACS) which is able to observe the operator's control actions, learn the task at hand, and take over a portion of the control responsibility. The computer functions not just as an intermediary between man and machine, but also as an intelligent participant in the control decisions and actions.

The basic decision made by the system is to select positions in space which are destinations of the manipulator end point. A single level decision network was used in the early work. Complete motion trajectories were not available, and fine control was left to the operator.

This paper describes the expansion of the concept to a multifunction learning structure which has the capability to generate a trajectory of motion and the sequence of fine hand motions which are required to accomplish a goal. The learning function is provided through two decision networks: trajectory control and sensory control. Trajectory control determines the translation pattern of the manipulator, while the sensory control function involves the generation of sequences of hand motions required to complete a handling operation.

As with the earlier single function learning control, the system is connected in parallel to the human operator, observing his control actions and developing an autonomous decision making policy (Figure 1). The manipulator can be controlled through direct operator control or through the lower ACS control loops under operator supervion. However, the multifunction system does not rely entirely on operator induced "training", since it contains some degree of capability for unsupervised learning. This permits certain autonomous adjustments of the decision policy to a dynamically changing environment.

The goal is to implement a control system that can be pretrained to perform a set of tasks such as search and identification of objects (in space and undersea) and industrial type packaging and assembly operations. For example, the system could learn to search for a particular type of object (e.g. geological samples of a particular size and weight), learn to discriminate such objects from others it encounters, and learn the sequence of pickup and handling motions involved in object collection. Such a machine will be able to adapt the repertoire to unexpected or changed local conditions – using remote supervision if necessary.

General Concepts

In the multifunction learning system the formal technique of describing manipulation in terms of state transitions is used in conjunction with a capacity for learning and probabilistic search. Central to the approach is the contention that transitional patterns can be learned from direct observation of human operator control over a manipulator. The system structure is based on a decision network whose decision space consists of all possible transitions between states. Given a specific manipulator state, an environmental state, and a task goal, the decision network selects the best transition to a subsequent state. The decision policy of the network determines the transition strategy. Decision policy is established by on-line parameter adjustment of the decision network to match the control policy of the operator.

The state transition structure is based upon earlier work which describes how remote manipulation can be defined in terms of distinct variations in the states of the remote manipulator and the states of the environment of operation (Whitney, 1971). For example position variation movement of object from one location to another, and opening or closing the claw are all changes in state. A task can be defined as a series of state transitions which forms a path between the initial and final state. Earlier manipulator control used heuristic programs for selecting a sequence of transitions between states that could accomplish task goals. Various examples of control programs for object manipulation and recognition which use this approach have been realized. These techniques have been mainly demonstrated in a "children's world" of block pickup and simple translation tasks (Ivancevic, 1971; Goto et al, 1972).

When the environment of operation includes complex objects, irregular, shapes, and more detailed tasks the application of heuristic programs becomes highly cumbersome. The number of possible transitions increases significantly and it is hard to predict all the allowed transitions. By using learning in a combination with a formal structure for combined state transition it is possible to obtain complex behaviour without establishing all the transition rules for a given environment and task. This has been explained earlier by Freedy, (1971). In the learning control system described in that report the role of the human operator was to supervise the Autonomous Control Subsystem. Faced with a new environment or unexpected circumstances the operator would be required to tqke active control of the manipulator. It is proposed here that by adding sensory inputs and a goal evaluation mechanisms the learning subsystem can lead the manipulator in a probabilistic search and further relieve the operator of active control responsibility.

Probabilistic search is used for autonomously adapting the transition patterns of the network to variations in the operational environment. When patterns of transitions which have been learned from the operator fail to progress toward a goal, the system selects alternate transitions until a goal is achieved. The system contains an evaluation function that recognizes the success of failure of a transition, depending upon the goal. This function provides an additional input to the decision network such that each transition is selected also on the basis of success or failure in an earlier transition as well task goal and manipulator state.

In this mode the network is continuously driven to make decisions regarding a transition. As successful transitions occur, the decision parameters which are associated with the given input data are reinforced, so unsupervised learning occurs. Since sensory data is most significant in evaluating goal accomplishment, the new learning element is called the Sensor Control Network. The Sensor Control Network and a Trajectory Control Network, an updated learning mechanism for generating gross manipulator motion, are described in the following section.

Sensor And Trajectory Control

The decision making mechanism for selecting future transitions is based on the maximum likelihood decision network of the type used in the early systems (Freedy, et al., 1971). Decision policy is determined by conditional probability parameters, P_{ij} . Those parameters are organized in an $N \times K$ matrix containing all the elements which relate N input events to K decision outcomes.

In this system the decision space D is a set of specific manipulator commands which are subtasks such as move left, close hand, etc. An input data vector $\overline{X}$, where $\overline{X} = (x_1, x_2...x_n)$, is used as the decision information input. The components of $\overline{X}$ represent input events which are relevant to the selection of an element of D. Decision input data is acquired solely from the following sources: (1) arm position information, (2) hand contact and touch sensor, and (3) operator inputs and commands.

The system contains two decision networks; one for the sensor control function and the other for the trajectory control. Each network has a corresponding input vector and an assigned decision space.

Sensor Control

The function of this portion of the learning subsystem is to employ sensory type information to recognize the achievement of a goal, then use that knowledge to reshape future decision policy and to trigger further autonomous action if necessary to achieve the goal. Such self-driven systems should learn to execute any task if:

(1) The decision space contains a set of subtasks which will accomplish the stated goal.

(2) The system has an evaluation function and sufficient input data to determine success or failure after each subtask.

The function of the decision network is simply to select a next subtask knowing the last subtask and its success or failure (represented by $S_{i,n}^s$ or $S_{i,n}^f$) manipulator status, sensory inputs, and the major task goal. The input vector $\overline{X}_n$ to the decision network at the nth stage is:

$$\overline{X}_n = \begin{bmatrix} \text{Last subtask, success or failure} \\ \text{Manipulator status} \\ \text{Sensor pattern} \\ \text{Task goal} \end{bmatrix} \qquad (1)$$

The decision space consists of a set of subtasks;

$$\text{Decision Space} = \begin{bmatrix} S_1 \\ S_2 \\ \vdots \\ S_k \end{bmatrix} \qquad (2)$$

A subtask is selected using the conditional probability measure:

$$P_{ij} = P\left[S_{n+1} \mid \overline{X}_n\right] \qquad (3)$$

which is the conditional probability of the transition $S_{i,n} \rightarrow S_{j,n+1}$. A subtask is selected by computing the conditional probabilities for all subtasks S_j ($j = 1,2,.....n$) and selecting the subtask S_j with highest conditional probability. The procedure for computing P_{ij} is identical to the process used in the original paper (Freedy et al., 1971). With the assumption of independence, the conditional probability in (3) can be calculated from conditional probabilities relating a subtask to identical components of the input vector. As for example, for the sensory input component x_1 and the subtask S_j there will be a probability element $P(S_j \mid x_1)$.

Training of learning involves negative and positive rein-forcement of transitional conditional probability parameters which are associated with the successful accomplishment of consecutive subtasks. Initially, in the untrained network, transition between subtasks will be random. As successfull transitions between subtasks occur and are positively reinforced, transitional patterns for goal accomplishment are formed. The training process can be described as follows:

    — Each successful transiton $S_{i,n}^s \rightarrow S_{j,n+1}$ leads to reinforcement of the conditional probability parameters which relate to the given input data at the nth decision stage to the selection of the subtask $S_{j,n+1}$.

    — Transition patterns which involve a failure in a subtask and subsequent success by subtask $S_{j,n+1}$ lead to reinforcement of the conditional probability parameters which relate $S_{j,n}^f$ to the selection of $S_{j,n+1}$.

    — Transition between subtasks which do not lead toward the accomplishement of the goals are inhibited by reducing the corresponding conditional probability.

    Typical transition patterns are shown in Figure 2. The decision space at each stage is illustrated by nodes. Arrows represent the possible transitions with the associated conditional probabilities P_{ij} . These probabilities are computed (for a specific given goal) at the completion of each subtask. For example, after the execution of subtask S_k at the nth stage is completed, the network computes the conditional probabilities for a transition to the $(n+1)^{th}$ stage. After that, the probabilities for $(n+2)^{th}$ transitions are calculated and so on. Transitional patterns are formed, through training, for the best sequence of subtasks required to accomplish a goal.

    The functional organization of the sensor control system is shown in Figure 3. In block d, the operation of each of the executed transitions is evaluated, i.e. whether the subgoal was accomplished. Evaluation of subtask success is done by a heuristic rule. For example, after a grasp subtask, the evaluation function checks whether the hand grasp sensor is on. If the subgoal is not accomplished, the network is forced to make a new decision and select another subtask. Continuous failure will force the system to continue and search for new transitions until the goal is accomplished. While the process of search continues, the decision network parameters are being estimated and unsupervised learning occurs.

Trajectory Control

    The function of the trajectory control network is to specify a unique set of points in space through which the manipulator end point must pass in transition toward a destination. The technique utilizes the learning properties of the computer system to learn typical trajectories generated by the operator.

The motion space of the manipulator is broken into discrete cells. The trajectory of movement of the arm is broken into a set of movement segments which link the cells together. Each segment is characterized by its direction rather than by its end points. In directing the hand to a destination point, the network performs a sequence of decisions to determine the instantaneous direction of the movement trajectory.

The decision space can be represented by a direction vector (d_x, d_y, d_z), where d_x, d_y, and d_z take any of the values $[+1, 0, -1]$. The decision input data are the present arm position, the past arm position and the current state of the hand.

The conditional probability parameters P_{ij} are defined as:

$$P_{ij} = P \text{ (output direction, } D_i \,|\, \text{Input events, } x_j \text{)}$$

where the input event is the present or past position or the state of the hand.

The number of probability elements conditioned on position grows quite large for finely resolved work space. To handle this problem the decision network contains a floating probability set. Probabilities of direction are recorded only at points where typical trajectories significantly change direction. Also, old probability values are discarded if they are not used in current decision making.

System Organization

Figure 4 illustrates an overview of the complete system. Control commands are generated by the trajectory and sensor networks. Both networks rely on the operator for major goal command input. Other decision inputs are obtained from the hand sensors and manipulator joint positions. In the training mode the operator generates his input goal command, but controls the task manually. At this stage transitional patterns are established.

For remote applications the system can be pretrained in a simulated environment under an operator's supervision. In the actual remote environment the system will operate autonomously. When unexpected changes in the environment occur, the system will search for alternate transitional patterns and reinforce the patterns which accomplish the goal under unsupervised learning.

Conclusion

The preceding sections have shown that it is possible to utilize a set of interconnected learning networks to provide a structure for nearly autonomous control of manipulator translation, fine handling motion and reflex control. The overall system makes up a close machine equivalent to perceptual motor coordination. It was also shown that by giving a learning network a set of criteria functions which correspond to its decision space, it is possible to obtain a search capability for required subtasks and to establish unique patterns of subtasks for major goal accomplishment. Unsupervised learning reduces the search time and provides for autonomous adjustments to dynamically varying environments.

In the main, the learning system offers a number of fundamental advantages; these are:

Communication between the supervising operator and the manipulator control system is performed naturally and efficiently through the standard controller. Instructions for changing behavior consist only of "showing how to do something." The need for a complex language covering continuous control commands is eliminated. Symbolic commands are restricted to major goals.

Performance Improvement is a natural concommitant of experience. In the operational environment correct responses are continuously reinforced, so that the number of correct decisions increases and errors decrease. Similarly, the general control strategy of the learning system is also improved. The overall result is a reduction in the amount of information transfer between the operator and the control system.

Reduced Probability of Error exists whenever a remote intelligence system is used. In the case of the learning machine the probability of error can be reduced by increasing the amount of pretraining. If sufficient training is given prior to the mission, erroneous decisions will occur only when there is an unpredictable change in the operating environment. In this case, the option of retraining remains and the system can be adapted to the new environment. Retraining is the performed through direct manual control. This is in contrast to the preprogrammed heuristic approach where unforseen environmental conditions may require a reprogramming process and a need for establishing new control algorithms. This involves off-line analysis, program testing and transmission.

In summary, it seems to us that further research in the application of learning systems to remote manipulation is highly worthwhile and is likely to yield significant results within practicable time spans.

REFERENCES

[1]   Ivancevic, Nebojsa J. and Luigi Cordella, "Artificial Tactile Perception with Computer Processing„ Proceedings of the Fourth International Symposium on External Control of Human Extremities, Dubrovnik, Yugo., Aug. 28 - Sept. 2, 1972.

[2]   Goto, Tatsuo, Kiyoo Takeyasu, Tadao Inuyama, Raiji Shimomura."Compact Packaging by Robot with Tactile Sensors„ Proceedings of the Second International Symposium on Industrial Robots, Chicago, Illinois, 149-159, 1972.

[3]   Whitney, D.E. "State Space Models of Remote Manipulation Tasks„ Ph.D. dissertation , M.I.T., 1968.

[4]   Freedy, A., F.C. Hull, L.F. Lucaccini, and J. Lyman, "A computer-based Learning System for Remote Manipulator Control„ IEEE Transactions on Systems, Man and Cybernetics, SMC - 1:356-363, October, 1971.

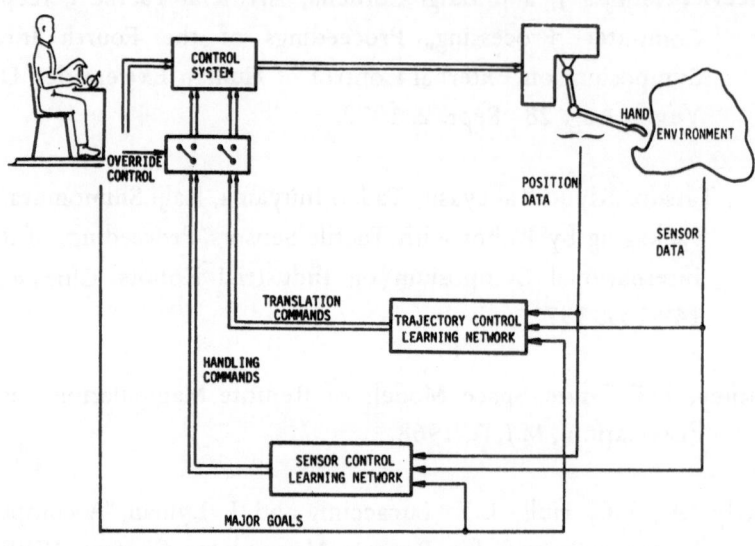

Fig. 1 Multifunctional learning system for manipulator control.

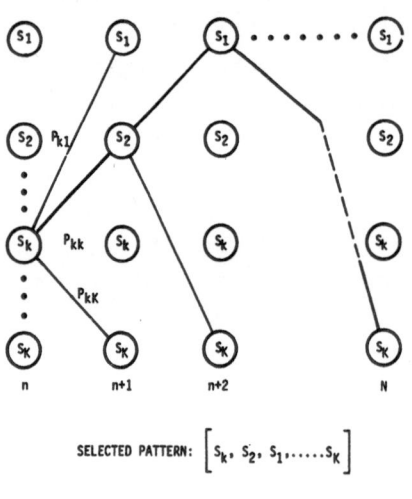

SELECTED PATTERN: $\left[S_k, S_2, S_1 \ldots S_K \right]$

Fig. 2 Transition pattern

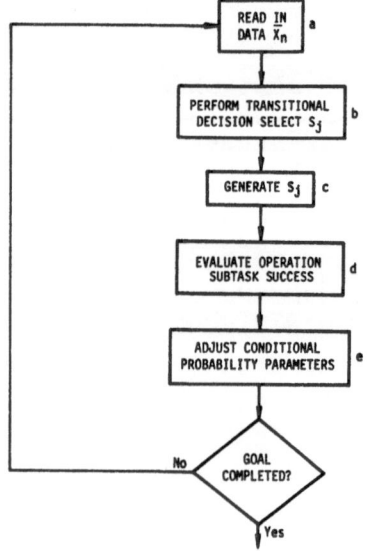

Fig. 3 Functional organization of
the sensor control system

ANALYSIS OF QUALITY OF MANIPULATOR MANUAL CONTROL

A.E. KOBRINSKII, Professor, Dr. of Engng Sci.
V.I. SERGEEV, Professor, Dr. of Engng Sci.
L.I. SLUTSKII, Post Graduate Student
L.I. TYVES, Cand. of Engng Sci.
Institute for the Study of Machines
Moscow, USSR

(*)

РЕЗЮМЕ

В докладе излагается методика, позволяющая подойти к формированию количественных оценок качества ручного управления манипулятором.Формируются критепии, включающие точностные и временные показатели работы системы "оператор-манипулятор".Проводится экспериментальное исследование,обсуждаются его рузультаты.

(*) All figures quoted in the text are at the end of the lecture

1. Introduction

The up-to-date manually controlled manipulators are characterized by a great number of mobility degrees, complexity of fulfilled tasks, and continuous intervention of a human operator whose task consists in composing the control programmes and realizing them. However, for estimation of quality of these highly functional and peculiar "man-machine" systems, as well as for quality estimates of other manual control systems, the subjective qualitative opinions of operators served as a basis up to now.

This paper concerns the methods which enable one to approach the formation of objective quantitative estimates of manual control quality. The methods are based on the idea that the quality of fulfilment of a great number of technological operations, associated with the use of manual labour, is characterized by two factors : accuracy and speed of their fulfilment [1].

In accordance with this premise the criteria are formed which include accuracy and time indices of "man-machine" system operation, experimental research is carried out, and its results are discussed.

2. Complex quality criteria

The basis of design of criteria (further on they will be called complex criteria) may be, according to [1, 2], the product of two values, one of which $F(T)$ is the time function of operation fulfilment T, and the second, $\Phi(\epsilon)$ — the accuracy characteristic (ϵ is the error at a certain time moment) :

(1) $$J = F(T) \cdot \Phi(\epsilon)$$

As for $F(T)$ it can be noted that its form depends on the weight given to this component of the complex criterion (1).

The accuracy characteristic $\Phi(\epsilon)$ is formed by comparison of the realized motion with the given (reference) motion, and the formation of the discrepancy of a certain index, for instance integral, modular or quadratic functional.

In particular, the accuracy characteristic may be a size of area between the reference trajectory and that realized by an operator. However, in practice it is more convenient to calculate an integral quadratic functional. Then, in the case of tracing of plane reference curve by the operator, the estimate of the accuracy characteristic is calculated by the formula

$$\Phi = \int_{\varphi_0}^{\varphi_1} (\rho - \rho_{_9})^2 \sqrt{\rho_{_9}^2 + \rho_{_9}'^2}\, d\varphi \tag{2}$$

where

$\rho_{_9}$ is the module of reference trajectory radius vector

ρ is the module of radius vector of the trajectory realized by the operator,

φ is the polar angle, $\rho_{_9}' = d\rho_{_9}/d\varphi$

The characteristic features of criterion of type (1) are its estimates given to the operator's work at boundary strategies of his actions. For instance, at maximum speed of operation and rather great errors, and when the work is very accurate and is done as slowly as possible such a criterion will have satisfactory values in spite of considerable values of error and task fulfilment time, respectively.

Such a situation can be avoided if, for estimation of manipulator manual control efficiency, the total criterion "time + error" is used [3]. Unlike the criterion of type (1) the above mentioned estimate is formed as a linear form of accuracy and type components

$$P = \Phi(\epsilon) + \lambda F(T),$$
$$0 \leqslant \lambda < \infty \tag{3}$$

The complexity of use of the complex criterion of this type consists in a possible incommensurability of its components. Attention was paid in [3] to the special importance of choice of the components values of the vector $W^0\{\Phi^0, F^0\}$ which are used in such cases for norming of criterion (3).

The major question, arising in connection with introduction of complex criteria, is as follows : whether they are sensitive to variation of system parameters, to what extent their estimate of system's quality change is differentiated, etc. In order to answer this question the system "operator-manipulator model" was studied with the use of an analog computer [1].

3. Experimental procedure

Below there are given the results of experimental study of systems of manipulator model manual control with the aid of complex criterion (1) in which $F(T) \equiv T$ is assumed, and the accuracy characteristic is calculated by formula (2). Realization of the boundary strategies does not satisfy limitations which regulate the fulfilment of the task by the operator.

With the aid of the analog computer of "A-110" type [4] and x-y recorder the manipulator output member and its drives have been simulated with a possibility of both astatic and positional control of its output component. The method of astatic control assumes setting up of motion speeds of the output member components. At the analog computer input there was installed a setting device of the model which is an x-y carriage with a handle, every coordinate of which is furnished with the motion pick up.

It was suggested to the operator to manipulate the handle and to trace with the output member of the manipulater model the reference trajectory with minimal possible error and time. The circuits for calculation of the criterion have been also realized on the basis of the analog computer. In the course of experiment the operators were interchanged, the manipulator model amplification factor was varied and reference trajectories were changed.

Three trajectories of equal length were taken as reference ones, Fig. 1. The equations of radius vectors of these trajectories are of the form :

$$\rho_{\ni I} = R_0,$$

(4)
$$\rho_{\ni II} = y_{02}|\sin\varphi| + \sqrt{R_2^2 - y_{02}^2 \cos^2\varphi}\ ,$$

$$\rho_{\ni III} = 2R_4 \sin\frac{\pi}{4}\ (|\sin\varphi| + |\cos\varphi|).$$

The choice of trajectories I, II, III is conditioned by the following reasons: a) comparative simplicity of setting-up of the given configurations on the analog computer. b) possibility of relatively simple estimation of complexity of each configuration [5].

According to the method suggested in [5] the estimation of complexity of a certain configuration is made by three characteristics: number of changes of curvature sign on the configuration (U), number of curvilinear sections of the configuration (V), and the sum of absolute increment of the inclination angle of the tangent (W). Using these characteristics let us introduce a generalized estimate of configuration complexity $Z = U + V + W/\pi$. Then, for reference trajectories I, II, III we obtain, respectively, the following estimates of their geometrical complexity: $Z_1 = 3; Z_2 = 9.5; Z_3 = 17$.

Specific features of computations on the analog computer leads, firstly, to the necessity of using the analytical expression of the derivatives ρ'_ϑ (see (21)) obtained by differentiation of equations (4), and, secondly, to the necessity of transfer from angle integration in (2) to time integration which is realized by the substitution

$$d\varphi = \dot{\varphi}d\tau \qquad (5)$$

4. Experimental results

The initial stage of the "operator-manipulator" system study was the stage of operator training. Prior to the beginning of recording the operation results the operator was given a possibility of doing one/two trial tracings in order to clear out the characteristic of the form of control under study. It was discovered that in the first several measurements the operator's error reached great values. However, further the values of error sharply decreased (a characteristic form of change of error in the function of the ordinal number of tracing n, which reflected the training process, is represented in Fig. 2). At the same time the improvement of skill of the operator takes place due to increase of tracing speed. Time variation of task fulfilment, as well as the change of the complex criterion (Fig. 2) in the process of training for the given complexity of the task, continues approximately during 25 or 30 cycles of tracing. The application of the positional control method does not change the nature of the above-mentioned curves.

Experiments were carried out further in order to study the relationships of values of the complex criterion of the "operator-manipulator" system operation quality and its components in changing the amplification factor of the x-y manipulator (under the amplification factor in the astatic control method with the high amplification factor the relation of the output member motion speed to the value of moving of the manipulator driving device is understood; in case of positional control the amplification factor may be considered as the scale one).

In the results of statistical handling of the experimental data it was found that in the astatic control method with great amplification factor it is natural for the operator to decrease the speed of tracing of the reference trajectory. Therefore, the dependence of task fulfilment time on the amplification factor is of the extremal nature (Fig. 3). The factor of the tracing speed decrease finds also its influence on the dependence of tracing accuracy on the amplification factor which has local extrema (Fig. 3). In this case the dependence of the complex criterion on the amplification factor is also of similar nature (Fig. 3). The abscissas of the graphs (Fig. 3) are the logarithms of the following values of the amplification factor η in 1/sec: 0.0675; 0.213; 0.675; 2.134; 6.75, each point on these graphs being the overaging result over ten experimental values.

In studies of manipulator positional control some changes have been introduced into the criterion calculation scheme stipulated by the technical complexity of calculating the value $\dot{\varphi}$ (5). The quality of task fulfilment for the positional control method was estimated by the simplified integral criterion

$$\Phi_1 = \int_0^\tau (\rho - \rho_{_\ni}) \, d\tau$$

It turned out that for positional control the integral estimate Φ_1 and the tracing time T have in the first approximation respectively direct and inverse (Fig. 4) dependences on the model amplification factor.

Comparison of task fulfilment results by trained operators, made with the use of the criterion J, made it possible to estimate the skill of each of them. Fig. 5 shows histograms obtained on the basis of processing of results of reference trajectory tracing by three operators, 50 times each.

The study of the influence of geometrical complexity of the reference trajectory on the quality of tracing showed the increase of tracing time with the increase of curve complexity. At the same time one should pay attention to the fact that there is an absence of a distinct tendency in the changing of tracing accuracy depending on configuration complexity. For instance, in examining the graphs of Fig. 6, one can notice both decrease of accuracy with complication of trajectory of one operator and improvement of tracing results, under the same conditions, of another operator.

It is possible to compare these results with the above-discussed regularities of the training process of operators. In the analysis of training process it was revealed that it runs more intensively in task fulfilment time whereas the error of its fulfilment on the main part of training decreases inconsiderably. Similar conclusion can be also made with respect to complication of reference trajectories: operator keeps the tracing error on a certain constant (evidently nominal for him) level at the expense of increase of time required for this purpose. Thus the accuracy estimate in choosing the control strategies by the operator in the task of tracing the reference trajectory is a major factor.

5. Analysis method of the "operator-manipulator" system quality

The executed study permitted us to establish a satisfactory "sensitivity" of the introduced criteria to the change of quality of system components. At the same time, while applying the described procedure to a real manipulator with its complicated geometry and peculiar limitations in kinematic pairs, it is necessary to keep in mind the features of its working volume. (Fig. 7). The requirements of satisfactory operation of the system in all points of its volume becomes evident. Such requirements were taken into account for instance in [6, 7] where the method of volumes is suggested and a manipulator service factor is sought. In the problem under discussion determination of volume averaged quality estimates can be done in the following way:

Let us consider the solid angle serviced by the manipulator (spherical sector with section shaded in Fig. 7) into several sections by equidistant planes, each section being a circumference lying in one of the working planes of the manipulator. Projection of all the obtained circumferences onto the greatest one will yield a set of concentric circumferences on its plane; smaller sets may be obtained in other sections of the solid angle. Now, in tracing these circumferences as reference trajectories, it is possible to take a certain integral index which takes account of the quality of tracing of each of them as a generalized criterion of operating efficiency of the "operator-manipulator" system. In order that the results of tracing a small radius trajectories be entered in the criterion on equal terms with the results of large trajectories, the norming of the obtained estimates should be used. Taking into consideration the above, it is possible to take as the integral index of the "operator-manipulator" system operating effectiveness, in the given solid angle, the following quantity:

$$\mathcal{H} = \sum_{i=1}^{m} \frac{Q_i}{\ell_i} \tag{6}$$

where $m = n(n+1)/2$ is the number of traced trajectories, n is the even number of sections, Q is the value of the complex criterion in tracing of the i-th reference trajectory, ℓ_i is the length of this trajectory. It is supposed that the manipulator manual control system which is best in its integral index of quality $\mathcal{H}(6)$ in tracing the plane analogs of space trajectories, will also more satisfactorily realize arbirary motions in the indicated working volume.

REFERENCES

[1] KOBRINSKII A.E., SERGEEV V.I., SLUTSKII L.I., TYVES L.I."To Quality Estimation of Manual Control Systems„ Machinovedenie, 5, 1971.

[2] KALININ V.N. "Generalized Optimality Criteria in Optimal Control Problems„ Automation and Remote Control, 2, 1965.

[3] SLUTSKII L.I. "On Quality Criteria of Manipulator Manual Control„ Machinovedenie, 2, 1972.

[4] YUFLER G.Zh."New Type of a General Purpose Computer„ Proceedings of the II International IFAC Congress, 3, M., Nauka, 1961.

[5] GANZEN V.A. GRANOVSKAYA R.M., "On One Method of Quantitative Estimation of Configuration Complexity „ Collection "Problems of Engineering Psychology", Issue 2, L., 1965.

[6] VINOGRADOV I.B., KOBRINSKII A.E., STEPANENKO Yu. A., TYVES L.I. "Some Problems of Manipulator Theory„ Proceedings of the II International Congress on the Theory of Machines and Mechanisms. Warszawa, 1969.

[7] KOBRINSKII A.E."Volumetric Criteria in Service Analysis of Manipulators„ Advances in External Control of Human Extremities. Belgrade, 1970.

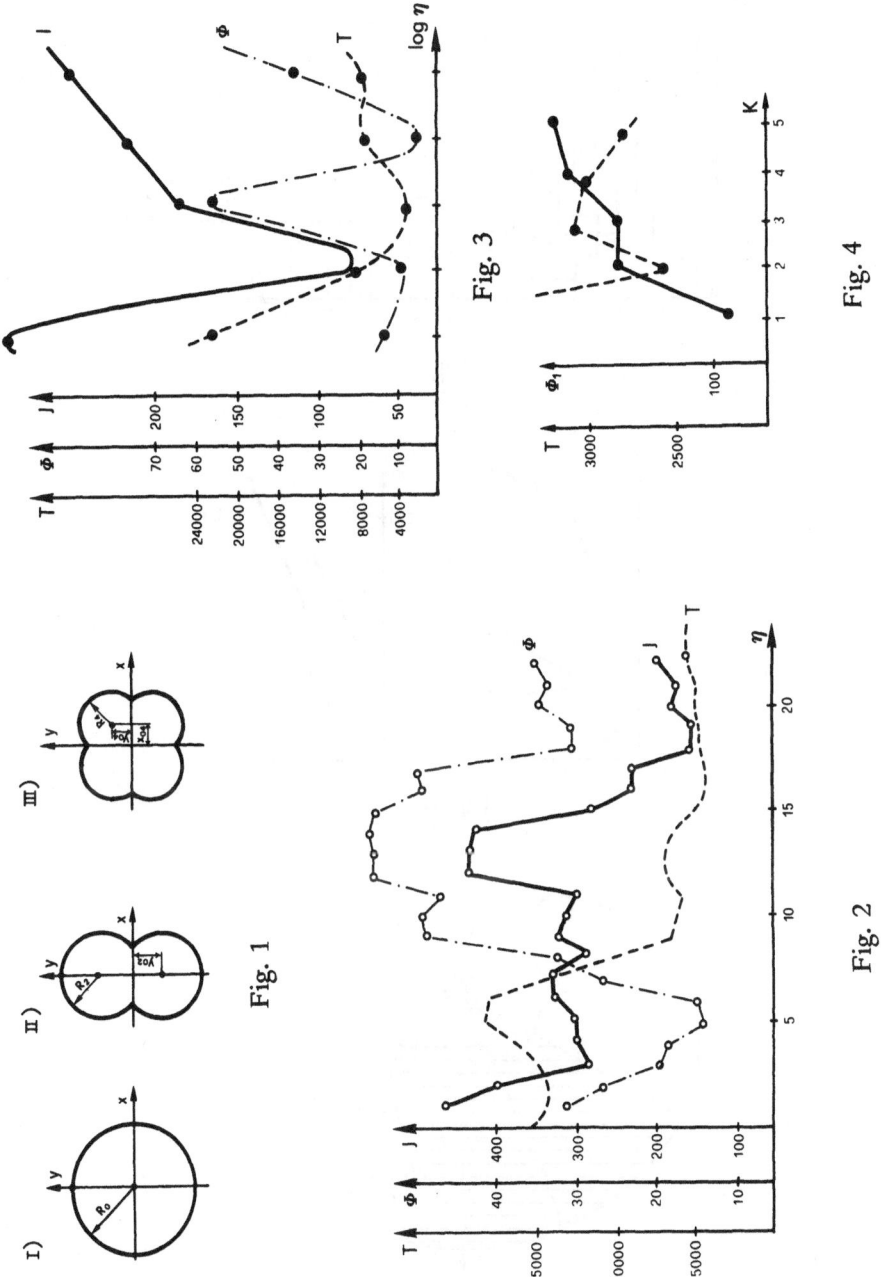

Fig. 1

Fig. 2

Fig. 3

Fig. 4

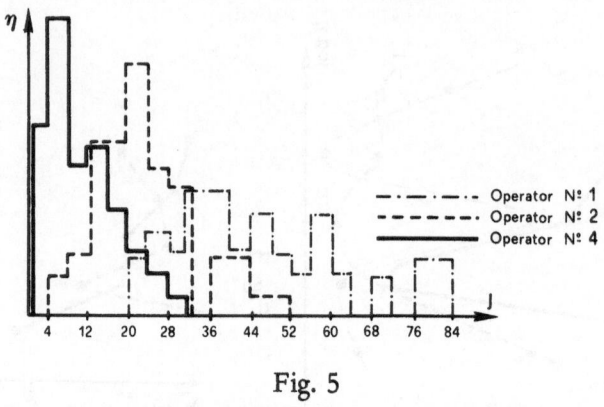

Fig. 5

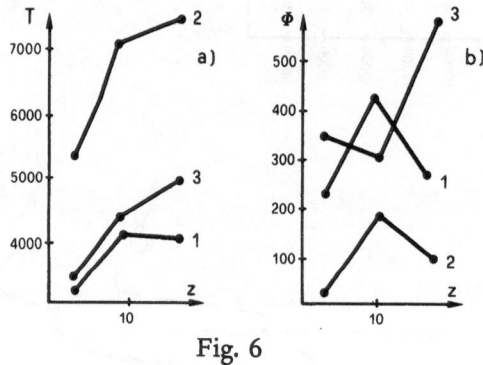

Fig. 6

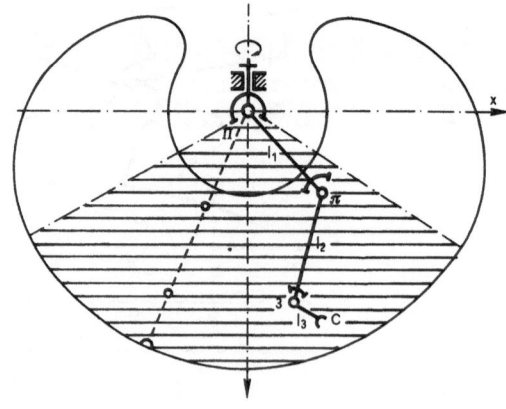

Fig. 7

MANIPULATORS FOR SUPPORTING AND SUBSTITUTING LOST FUNCTIONS OF HUMANS EXTREMITIES

Adam MORECKI, Professor
Insitute for Applied Mechanics
Warsaw Technical University, Warsaw, Poland

Zoigniew BUSKO, M.E.
Institute for Applied Mechanics
Warsaw Technical University, Warsaw, Poland

Henryk BOROWSKI, M.Sc.
Institute for Applied Mechanics
Warsaw Technical University, Warsaw, Poland

Kazimierz FIDELUS, Ass. Professor
Insitute for Biologic Sciences
Academy of Physical Education, Warsaw, Poland

Edmund KOTWICKI, M.E.
Institute for Applied Mechanics
Warsaw Technical University, Warsaw, Poland

Roman PASNICZEK, M.E.
Rehabilitation Centre
Medical Academy, Warsaw, Poland

Krzysztof TEMPINSKI, M.E.
Institute for Applied Mechanics
Warsaw Technical University, Warsaw, Poland

(*)

Summary

     The problem of assisting or substituting for lost functions of human extremities, by applying various manipulators, lies in the field of interest of modern biomechanics. This paper deals with some biomechanical, thechnical and medical problems encountered during a selection of type of device and a control system of such manipulators as orthosis, prosthesis and implant stimulator. A few examples of designs of such apparata are given.

(*) All figures quoted in the text are at the end of the lecture.

1. Introduction

The problem of assisting or substituting for lost functions of human extremities, by applying various manipulators, lies in the field of interest of biomechanics. This paper deals with some biomechanical, technical and medical problems encountered during a selection of type of a device and a control system of such manipulators as orthosis, prosthesis and implant stimulator. A few examples of designs of such apparata are given.

Very often, proposed devices have to meet contradictory requirements. The more important of them are listed below:
— functionality as high as possible, which calls for a design of a performing system with as great a number of optimum selected degrees of freedom as possible, optimum range of movement, reliable and hygienic connexion with the patient's body,
— as physiological, reliable and efficient a control system as possible, which needs minimum control sites and leads to automation of movement procedures,
— application of high capacity supply resources,
— maintaining the high efficiency of patient during long periods of using a manipulator (elimination of fatigue and discouragement) and developing a conviction that the device is indispensable in every-day activities.

The paper describes procedures leading to optimum solutions that are a conscious compromise between the needs and the physiological and technical possibilities of the patient.

2. Biomechanical assumptions

The loss of a function of an extremity may result from the damage of the central nervous system, amputation or less frequently from the atrophy of muscles (drives). Amputees are equiped with prostheses (prosthetic manipulators) and other patient's cases are equipped with orthoses (orthotic manipulators). In cases of cord lesion, movement of an extremity may be evoked by means of nerve or muscle stimulation, the latter method being used also in case of damage of peripheral nervous system. This kind of stimulation may be carried out by means of implant stimulators.

As a rule, a manipulator has smaller number of degrees of freedom than a human extremity. Its end has one, less frequently two degrees of freedom (instead of 23) used for prehension. In order to determine the significance of the function to be assisted or substituted for, one can use the results of analysis made for areas being within the reach of the arm (1, 2). Here is a list of movements arranged according to the magnitude of the useful area: flexion and extension in the elbow joint, suppination and pronation of the forearm, flexion and extension in the wrist joint and movements of the shoulder joint. The latter may be partly compensated for by movements of the shoulder girdle and the torso.

The order of the amplitudes of the movements of a manipulator should be the same as in case of the human extremity if the manipulator has to reach out to proximity of the head and face. Values of the torques developed in the joints should exceed of least twice the maximum value of the torques resulting from the gravity forces generated during the working movements. These values should be increased if the manipulator is intended for lifting loads.

If the manipulator is to be carried about by the patient the designer should take into account proportions (length and width) and adequate moments of inertia of individual members of the human extremity.

Very important is a selection of actuators which should work in a similar way to muscles.

3. Selection of actuators and control system

A comparative analysis of pneumatic, hydraulic and electric drives shows unquestionable advantages of the former. The most important advantages are:
– general possibility of simulating basic properties of the skeletal muscle,
– small over-all dimensions and weight in relation to the output power,
– good controllability,
– reliability and long durability,
– noiseless operation,
– lack of vibration during operation,
– low energy consumption
– relatively low manufacturing costs.

A comparative analysis of pneumatic, hydraulic and electric drives, made from the standpoint of contradictory requirements imposed on designs of prostheses and orthostheses, shows that in 10 cases out of 19 pneumatic drive is used. This is why this type of drive was selected for developing medical manipulators.

The number of the selected movements to be assisted must be limited for technical and physiological reasons.

It is known from experience that the number of movements greater than 3 - 4 inflicts great difficulties upon the designer of a control system, which needs excessive number of control sites and more complex supply resources. For this reason the authors limited the number of movements of the assisting device to 3 and for devices substituting for the lost human extremity to our movements. The maximum number of the control sites was assumed to be one or two, at the most, and the movements to be performed one after another.

4. Portable orthotic manipulator

After analysing various pneumatic actuators, the authors choose a plunger-cylinder type for elbow flexion, pronation and suppination, and for prehension the bellows type was chosen.

A system having 3 degrees of freedom needs a control system with 6 outputs.

It is known from practice with orthotic and prosthetic manipulators, controlled bioelectrically it is very disconcerting for the patient if some of his muscles function as control sites. Taking into account two contradictive terms, i.e. minimum number of control sites and desirable simultaneity of movements, the authors considered it better to limit the number of control sites at the cost of performing movements one after another. In this case one muscle may be used for cooperation with the two-level control system which is shown in Fig. 1.

The 1st (higher) level of the signal selects the type of movement to be performed; the second switching on of the same signal changes the type of a movement. The 2nd (lower) level of the signal is used for performing a selected movement. The first switching on of this signal results in performing a selected movement while the second switching results in performing an antagonistic movement. Time constants are selected in a way which permits a desired movement to be performed without performing an antagonistic movement. Bioelectric signals are amplified and adequately transformed in a logic system and control a set of electro-pneumatic valves operating the inlets of the actuators for the compressed air.

A diagram of a logic system is shown in Fig. 2. This system was built from HIL (high level logic) elements. The system consists of the snap-action switches and NOR elements.

The block diagram of an orthotic manipulator is shown in Fig. 3. The amplifier of myopotentials, logic system, and electropneumatic valves are supplied from 12 V battery; the valves are supplied from a gas cylinder containing liquid CO_2. The drives and control system have been tested for durability and reliability with good results.

5.Stationary orthotic manipulator mounted on the wheel-chair

This device is intended for assisting the lost functions of upper extremities in case of $C_5 - C_6$ lesion of the spinal cord. This ailment practically eliminates prehension and considerably impairs elbow and shoulder movements. Only the patient's head retains previous possibilities of movements which are utilized for controlling movements of both extremities.

Taking advantages of the assumptions described in paragraph 2, an assisting system having 6-degrees of freedom for each extremity has been developed, ensuring full range of movements in elbow and shoulder joints as well as prehension. In this system wrist movements as well as movements of individual fingers are not assisted. The structural scheme of the orthotic manipulator is shown in Fig. 4.

It is an open spatial mechanism having 6 degrees of freedom, composed of S kinematic pairs of 5th class. A double kinematic pair of 5th class with axes of rotation 0_1 and 0_2 makes it possible to move the apparatus and extremity in relation to the torso, so that the patient is able to assume the most comfortable position. The extremity fixed to the apparatus, has 6 degrees of freedom. A symmetrical system will be provided for the other extremity. As the need arises, the patient may use only some of the degrees of freedom. The design of the splint 4, of the mechanism R_6 for prehension, and R_5 for pronation and suppination of the forearm is identical (the splint 2) as the portable design. Elbow flexion (in the joint R_4) is performed by means of a special pneumatic actuator, constancted as a long stroke cylinder with a flexible diaphragm. Shoulder movements are performed by means of the splint 1, artificial pneumatic muscles and angular bellows (R_1, R_2 and R_3).

Though the ranges of movements for individual joints of the manipulator are smaller than for normal human extremity, they enable the patient to carry on every-day activities.

The whole structure of the manipulator is mounted on the electrically powered wheel-chair which enables the patient to move around his flat and relieves him from carrying the load of the manipulator. This control system does not constrain the patient and enables him to change his position whenever he wants to. The block diagram of the control system is shown in Fig. 5. Designers, while selecting a control system, put emphasis on simplicity of the system (simple in operation and non tiring for the patient) and its reliability. For this reason the system has been made from simple two-positional micro-switches. Eight micro-switches (B) are placed circularily so that the patient can operate them with slight movements of his chin placed in a shallow socket. Eight directions of movements of the patient's chin are distinguished: forward-backward, left-right and middle directions 1-6. Direction T is designed for selecting one of the extremities (left or right) to be controlled. The selection is made by sequential pushes exerted on the lever D. After selecting an extremity, the patient selects a proper movement (for example flexing or extensing) by pushing the lever in direction P opposite to T.

     After selecting the extremity and the direction of the movement for example: left extremity, flexing in all joints (this state being signalized by special lamps), the patient may proceed with further controlling. By pressing the control lever D in the remaining 6 directions, the patient starts the desired program for a movement. Twenty-four electropneumatic valves P_1 to P_{24} , two for each actuator, are used for inflating and exhausting actuators. The range of movements can be changed by varying the time of inflation. This time is long enough to ensure continuous controlling. The control system is twopositional (0/L). Any of 24 electro-pneumatic transducers P_1 to P_{24} may be put into action if a proper signal z_1 to z_{24} is of value L. If so, an actuator performs corresponding movement 1. Signal value z = L is obtained as a result of a conjunction of signals s = L, x = L, y = L.

     If the control lever is pressed again in the direction T, the snap-action switch 1P is put one of the states 0-L. Signal 0 stands here for controlling the right extremity, signal h - for the left extremity. Switch P operates in a similar way: signal 0 means inflation of the actuators and signal L means exhaustion of the actuators. Signals y_1 and y_2 are negations of signals x_1 and x_2.

     The table in Fig. 5 shows the values of individual signals in the case when the patient presses the micro-switch in direction 1 (elbow movement) and switches 1P and 2P are set in position 0. As a result of this an elbow-flexion of the right extremity is performed. The above control system for extremities, may be very easily accomodated to controlling the wheel-chair ride, by mere pressing the lever down with the chain. This seemingly complex control system can be mastered by the patient, in relatively short time, with a proper training. In spite of sequential performing of movements, this control system ensures precise positioning of the extremity with no need for a computer which is indispensable in the case of a continuous and simultaneous controlling system with 12-degrees of freedom.

6. Prosthetic manipulator of forearm with hand

After high-level amputations, above the elbow, it is necessary to assist the functions of the last part of the extremity by equipping the patient with a manipulator. The remaining part of the extremity, in case of unimpared shoulder joint, retains only 3 degrees of freedom out of the natural 30. Reconstruction of a system with so many degrees of freedom is very difficult from the standpoint of controlling. Biomechanical analysis aimed at putting the movements in order according to their servicing volume, enabled researchers to select the minimum number of movements indispensable for a prosthetic manipulator. They are as follows: prehension, elbow flexion and extension, forearm pronation and suppination and wrist flexion and extension. On the whole a manipulator ought to have 4 degrees of freedom. Pneumatic actuators of bilateral action and electro-pneumatic valves, similar to those used in the orthotic manipulator, have been used in driving system of the manipulator, the design of the prosthetic manipulator is shown in Fig. 6. Elbow and wrist flexion of extension and finger movements (mainly 2nd and 3rd fingers), since the thumb is set passively are performed by means of independent pneumatic actuators (Fig. 6) and unilateral lever. Rotation of the forearm is performed by means of an independent actuator and tapped sleeve with a spline.

The block diagrams of the prosthetic manipulator, control and supply systems are shown in Fig. 7. The control system is of a bioelectric type with 2-3 bioelectric signals for sequential controlling the eight active movements. The pneumatic actuators are supplied from a gas cylinder containing liquid CO_2; the control system is supplied from 12 V storage battery.

7. Returning the lost movements to the extremities by means of implant stimulators

Stimulation of muscles or nerves by means of implant stimulators is a new method of returning the lost functions to the extremities. Particular applications may be found in case of paralysis of all four extremities which results from lesion of the spinal cord on the neck section (tetraplegy). In this case not only muscles of the lower extremities are paralysed, but also muscles of the forearm and fingers. For this reason the patient is not able to sit and carry on daily activities. Unfortunately we have no means to regenerate the damaged section of the spinal cord and to restore anatomical and functional continuity of the structure within the central nervous system. However, experiments with electric stimulating of paralysed muscles and nerves are more and more frequent. Their purpose is to restore some of the lost functions.

Before implanting stimulators, the patients (tetraplegics) were put to a series of tests, their purpose being to determine the effect of stimulation on the paralysed muscles. The results have so far been satisfactory. It has been found that the external stimulation of the nerves prevents the muscles of the paralysed extremities from fast atrophy and no side-effects have been detected after 15 months.

The first experiments with direct muscle stimulation were carried out 3 years ago. For the next 7 cases we applied the nerve stimulation, implanting 13 stimulators. The aim was to excite the muscles directly and to make use of the aferential signals.

The first clinical approach was to determine whetehr the implant stimulators are well tolerated in the body. Investigation was also made to determine the effect of systematic stimulation on the muscle bulk, developed forces, and time of maintaining maximum moments of forces developed by stimulating individual muscles or sets of muscles.

The last 4 cases have been carefully investigated from the standpoint of active influence of the patient on the results of the stimulation and the effect of the training on the improvement of the rudimentary active mobility of the hand. In these 4 cases 9 stimulators have been implanted, mainly on the medians nerve.

The surgery was performed under local anesthetization which enabled the researchers to evaluate the mid-operative effect of the stimulation and find the best position for electrodes so that the stimulation was the most effective and least painful.

Systematic training was started a few days after the surgery when the wound had healed up and the stitches had been taken off. The block diagram of the stimulating system under discussion is shown in Fig. 8. Grasping force was measured by means of a special beam with strain gauges glued to it, and an amplifier. Simultaneously, EMG signal was used for stimulating muscles by means of a stimulator affixed to the n.medianus. The stimulator was controlled by a programming device which switched it on and off according to the present programme. The training was carried out every day according to the following programme: The stimulator was switched on intermittently for periods of 5 s. with 30 s. intervals. The amplitude of the stimuli was so adjusted that the maximum tetanus of finger flexors and oppositor of thumb occurred. This cycle (stim-ulation-rest) was repeated for 5 minutes and then the stimulator was switched off for 55 min. and then switched on again for 5 min. of stimulation, and so on. Daily workout for the patient lasted from 2 to 4 hours (3-5 stimulation sessions, 5 minutes each). The time of duration of individual impulses and their frequency were as follows: duration time of the impulse - 0,7 ms, frequency - 15 cps. Occasional adjustments for these parameters were made for individual patients. After applying this method of stimulation we observed in all cases a gradual increase of forces developed by muscles (Fig. 9) and increase in muscle bulk, which was apparent from a slight increase of circumferences of the extremities.

All the patients who underwent the training were able to develop the maximum force for a longer time. The maximum moment developed by the thumb's flexor and the finger opposing the thumb could be increased up to 18-20 s as compared to 10-12 s before the training. The first results were given in [4]. Further investigation confirmed this conclusion and showed that in result of the stimulation the patient was able to increase voluntarily the values of forces developed by paralysed muscles.

In some cases the effort made by the patient brought negative results when a developed force decreased during the experiment. Fig. 10 shows a record of an increasing force developed with active assistance of the patient. Fig. 11 shows the case with negative results. In these figures A means the starting point of the stimulation, B means the starting point of active assistance of the patient, C means the end of assistance, and D means the end of stimulation. It was also found that both effects depend on the level of stimulation (stimulation back-ground).

The records of bioelectrical activities of m. flexors and the finger opposing the thumb are faithful representation of mechanogram. Active assistance of the contraction with positive effect is represented in EMG record by an increase of amplitude and synchronization and enrichment of the record. The negative effect of assistance is followed by de-synchronization, weakening of the record and drop of the amplitude of potentials. It is necessary for a patient to have a proper control system in order to use functionally the stimulator. This control system ought to enable the patient to switch the stimulator on or off, without help of other people.

The control system should meet the following requirements
— it should not constrain the natural movements of the patient, especially the movements of his head and upper extremities,
— the system should enable the patient to ensure an adequate amplitude of stimulating impulses. The system should memorize this amplitude and reproduce it at request many times,

- the system should enable the patient to switch it on and off, independently,
- the system should be simple in service, reliable and not demanding a special training or effort on the part of the user,
- switching the system should not engage the muscles or movements of extremities necessary for performing other activities.

Various possibilities were considered during the preparatory works as far as the indicators and control system were concerned. Out of many types of indicators special attention was given to micro-switches, mercuric two-positional indicators and acoustic indicators (for example various loudspeakers). Finally, the mercuric inclination indicators were decided on. For switching them, movements of the patient's head were utilized. The range of movements necessary for switching the system was so selected that random movements of the head did not affect the control system. The system can be switched on and off by forward, backward movements. The block diagram of the control system is shown in Fig. 12.

Four inclination indicators 1, 3, 2 and 4 are mounted on the frame of the spectacles. Signals from the indicators pass through the reley system P, and control the four-state system, which switches the stimulator transmitter and potentiometer system with memory on and off. Switching the system by means of the indicator 1 puts the stimulator into action and switches on the motor connected with the potentiometer, which in turn changes the amplitude of the stimulating impulses (Fig. 13). When the desired level of the amplitude is reached, the patient stops the motor by means of the indicator 3. Transmission ratio is so matched that the time of selecting the amplitude of the impulses is about 10 to 15 s, long enough for the patient to make a proper choice safely and without haste. The indicator 2 is used for switching the system off and for multiple switching with the pre-set amplitude (with no need for setting it every time). The indicator 4 switches the system off and sets it at the starting point. The motor is supplied from accumulators mounted on the wheel-chair. The above control system may be used by a sitting (in a wheel-chair) patient and, with some restrictions concerning the position of the indicators, by a reclining patient. On the basis of this investigation one can draw the conclusion that electric stimulation applied in tetraplegy-stops atrophy of muscles, and that systematic training increases the maximum value of forces developed by stimulated muscles.

Moreover, the stimulating impulses work their way through nervous system, which ensures a correct contraction of the paralysed muscle. Electric stimulation enables the patient to move with paralysed extremities and partly returns prehension.

Conclusions

The described designs of prosthetic and orthotic manipulators as well as implant stimulators are being investigated technically and clinically. The results obtained so far show that they are functional and work fairly well.

REFERENCES

[1]   A. MORECKI, J. EKIEL, F. FIDELUS - Bionika ruchu, PWN, 1971

[2]   A. MORECKI, K. FIDELUS - Niektore zagadnienia rehabilitacji funkcjon-
      alnej ruchów kończyn ludzkich, PROBL.TECH.MED., 1972, III Nr 3

[3]   A. MORECKI, K. FIDELUS, K. TEMPINSKI, M. DEWERT - Functional
      orthotic of upper extremities, The Fourth Inter.Symp. on External
      Control of Human Extremities, Dubrovnik, August 28 - Sept. 2 1972

[4]   R. PASNICZEK, J. KIWERSKI, J. WIRSKI, H. BOROWSKI - Some
      Problems of Implant Stimulation Applied to Grasp Motion - The
      Fourth Inter. Symp. on External Control of Human Extremities.
      Dubrovnik, August 28 - Sept. 2, 1972.

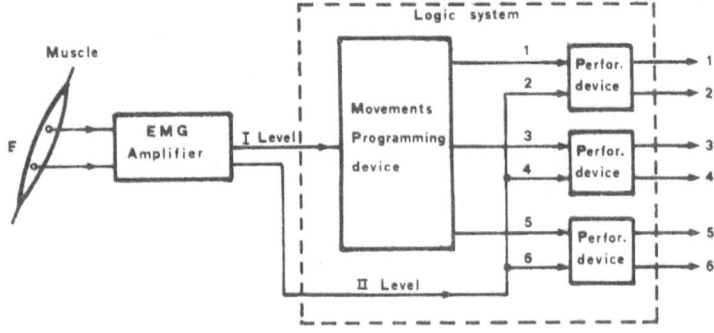

Fig. 1 Block diagram of control system

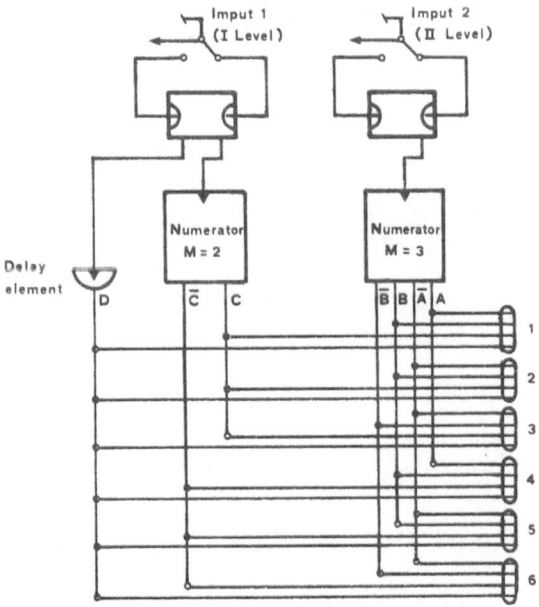

Fig. 2 Logic diagram of control system

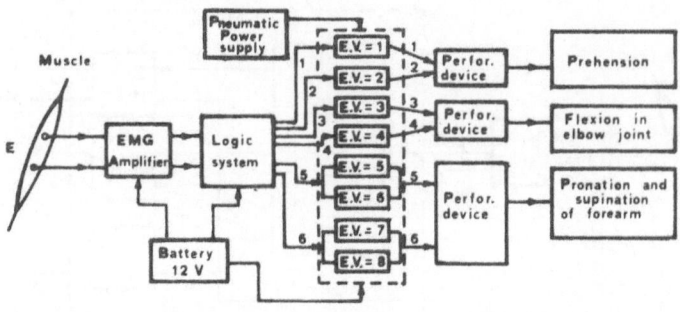

Fig. 3 Block diagram of 6 channel orthotic manipulator

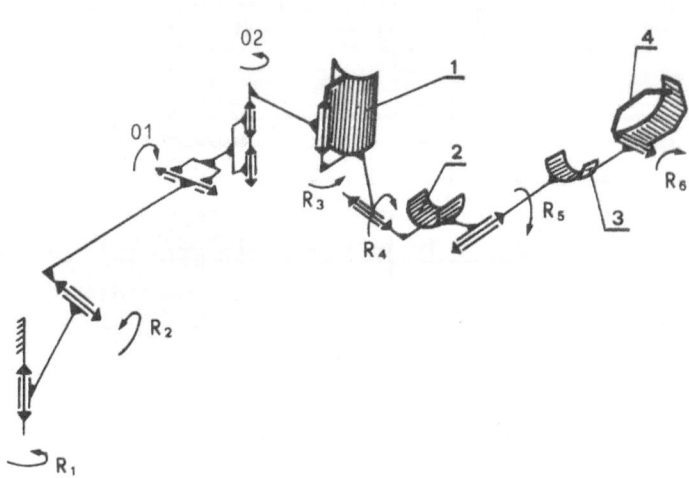

Fig. 4 Structural scheme of orthotic manipulator with 6 degrees of freedom

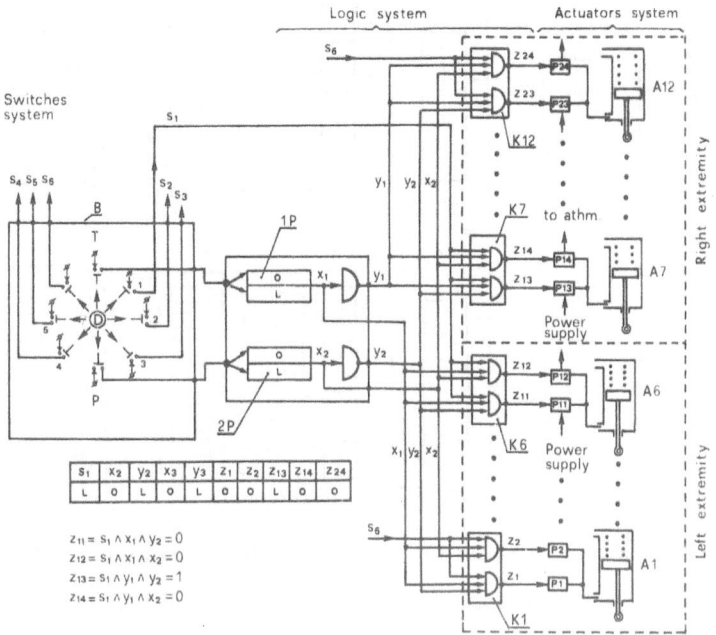

Fig. 5 Diagram of system of control for orthotic manipulator
mounted on a wheelchair

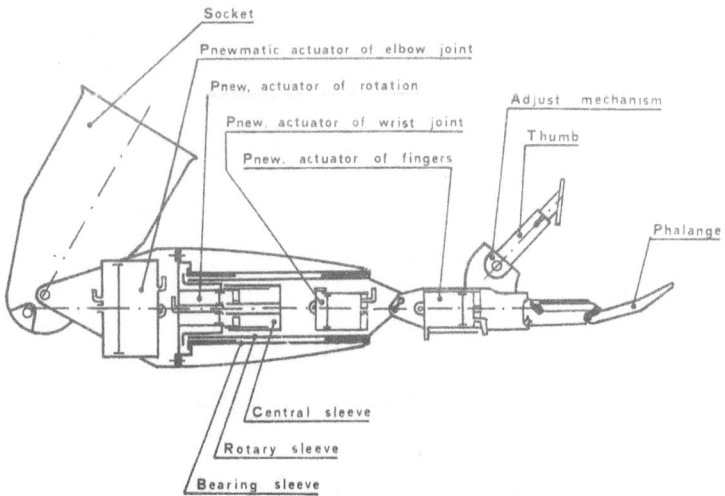

Fig. 6 Design of prothetic manipulator of upper extremity

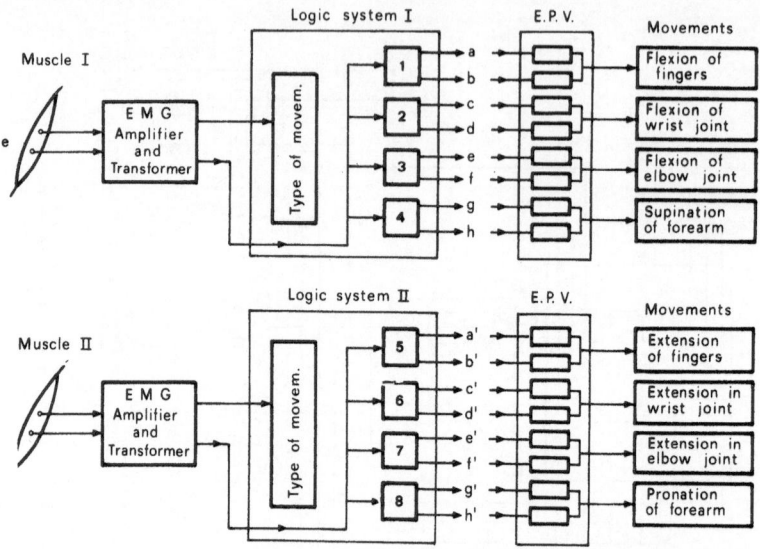

Fig. 7 Block diagram of 8 channel prothetic manipulator E.P.V.

electropneumatic valve

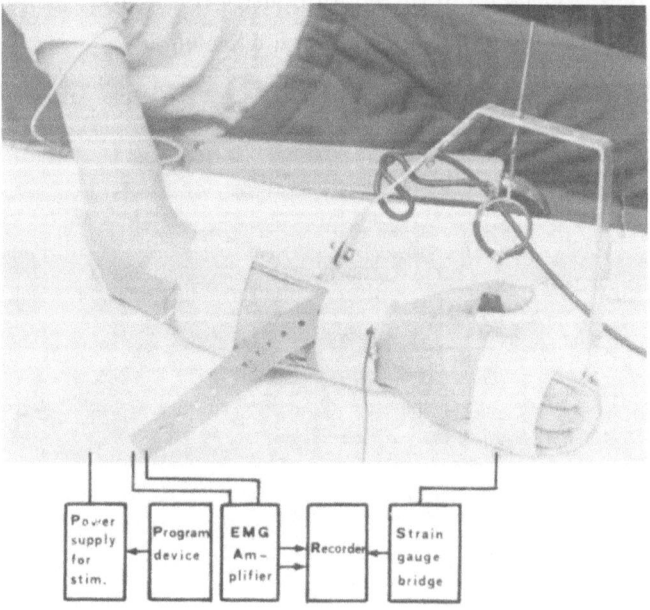

Fig. 8 Block diagram of stimulating and measuring system

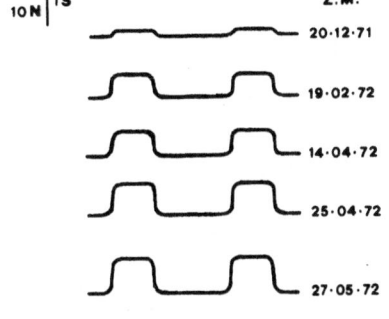

Fig. 9 Force increase after 5 months of stimulation

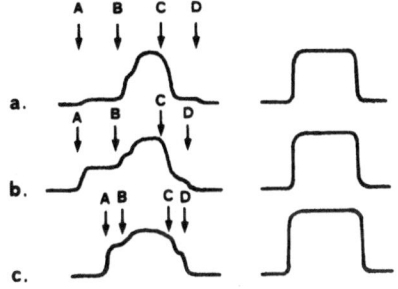

Fig. 10 Force increase during stimulation with initial level

a — first level

b — second level

c — third level

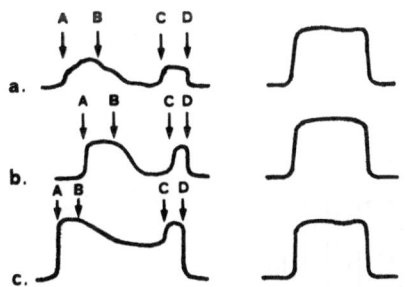

Fig. 11 Force decrease during stimulation with initial level

a — first level

b — second level

c — third level

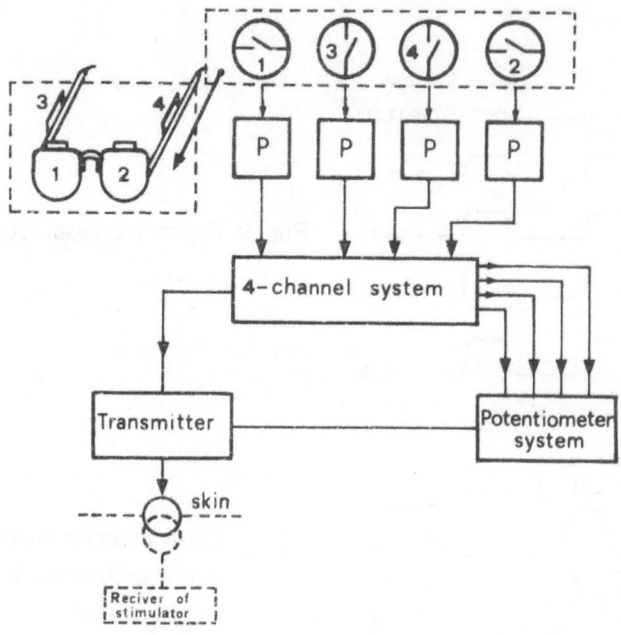

Fig. 12

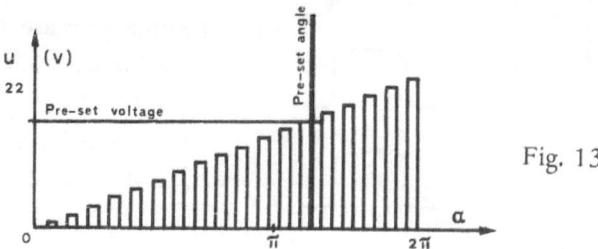

Fig. 13

DESIGN OF MASTER—SLAVE MANIPULATORS :
BIOTECHNICAL ASPECTS

E.P. POPOV, correspondent-member of the Academy
of Sciences, USSR, Moscow, USSR.

N.A. LAKOTA, Senior research worker of the Academy
of Sciences, USSR, Moscow, USSR.

(*)

Summary

  The paper is devoted to some biotechnical aspects of master-slave manipulators design and experimental studies of the operator-manipulator system. Classification of M/S manipulators and the formulating of general biotechnical problems was also done in the paper.

  The results of these experimental investigations, done on the physical models of M/S manipulators and on the anolog computer models made it possible to find a good design solution to the problem of designing M/S manipulator.

(*) All figures quoted in the text are at the end of the lecture

The master-slave manipulators are the most general-purpose technical systems enabling the performance of complex technological operations with great precision owing to the unambiguous correspondence between spatial position of the operator's hand and the grip of the manipulator slave. They may be classified by construction, control, operating conditions, etc. [1] .

The basic distinguishing feature of master-slave manipulators is the manner of control which defines, in the main, the structure and composition of systems required for the manipulator to have the necessary properties and performance. All kinds of master-slave manipulators (atomic, submarine, space, industrial) may be classified by the manner of control as follows [1] : directly controlled manipulators, remotely controlled manipulators, and automatically controlled manipulators.

By the type of slave actuators all manipulators may be broken into electrical, hydraulic, pneumatic, and mechanical. The choice of slave is defined by lifting power, application, and operating conditions of the manipulator.

The second distinguishing feature is the principle of master-slave manipulator operation. By this feature the manipulators are divided into unilateral and bilateral. In unilateral manipulators, slaves are controlled either by a human operator through a master unit, or, under automatic control, by a special control unit.

Control systems of unilateral manipulators may have force feedback or not. In the latter case they are conventional systems without force feedback, closed or opened by position. In particular, their features are defined by the specific object of control. Since in unilateral manipulators the force feedback is passive, their control systems are called "systems with passive feedback". In such manipulators the coupling between master and slave does not carry the load.

In bilateral manipulators, master and slave are controlled by torque applied both by the operator and the local. Such manipulators, characterized by the active force feedback, may be broken into two main categories depending on the kind of coupling between master and slave that may carry the load or not.

Bilateral load coupling manipulators (having either mechanical couplings, or synchronous tracking ones) are used only for insignificant distances between the operator and the object of operation [1] . Increase in their lifting power together with provision of high sensitivity are intimately related to the design of torque compensation and torque scalling systems.

The master-slave manipulators with active force feedback and non-load coupling between master and slave enable one to place the operator at a significant distance from the object of operation. Bilateral action of such manipulators is assured by symmetric control with active force feedback having dependent channels for position and fore tracking feedbacks, and non-symmetric ones having independent channels for position and force tracking feedback [1].

The former systems are widely used because of simpler implementation of sensing elements. Realization of high performance symmetric systems requires further development of the theory because they belong to a special class of two-dimensional automatic control systems. The non-symmetric systems assure high performance in tracking position and torque reproduction. However, their design is greatly encumbered by the technical imperfection of the load torque measurement methods.

The classification [1] adopted by the authors enables them to describe a wide class of master-slave manipulators by their distinguishing features, and to formulate a number of general biochemical problems arising in their design.

Here are several of them :

1. Design of masters and slaves assuring operation over the whole operation zone and preserving complete controllability of all degrees of freedom. These problems are closely related to the biomechanical and structural analysis of the construction of human hand.

2. Design of manipulator control systems and sensing devices. This is closely related to the study of the basic dynamic characteristics of the human arm (i.e. time of response, speed and precision of movements, power characteristics, and control of efforts), to the investigation of the visual perception channels, etc.

3. Study of the man (operator)-manipulator complex as a biotechnical system. These problems are related to the identification of a manipulator functional diagram and the manner of control assuring the maximal effectiveness of the biotechnical system.

The master-slave manipulators, as opposed to other types of manipulators, to a great extent should correspond to the anthropometric and physiological peculiarities of the human operator performing control functions. This necessitates the biomechanical analysis of the human arm including structural, kinematic, and dynamic study. The structural analysis maked it possible to examine the number of degrees of freedom of the human arm with due regard to peculiarities of its joints. The human arm may be represented by an open kinematic chain having twelve degrees of freedom (Fig. 1a). But for the operator's actions to be natural, it is sufficient to make the master and slave kinematic diagrams similar to that of the human arm beginning from the hand up to and including the elbow joint. Arbitrary spatial position of the elbow joint may be assured by a two-axial joint together with telescopic and rotational ones. In doing so, the number of degrees of freedom of the human arm kinematic diagram drops down to ten (Fig. 1b). Since the use of the telescopic joint in the shoulder is not always possible, the total number of degrees of freedom in the original human "arm" kinematic diagram may be reduced to nine. This diagram will be used for the determination of master and slave kinematics.

A.E. Kobrinskii and Yu.A. Stepanenko [3] were the first to carry out comparative examination of various kinematic diagrams for manipulator masters and slaves.

They have suggested criteria ("degree of manoeuvrability" and "service-factor") defining important features of manipulators.

Increase of manipulator effectiveness led to imposition of two more conditions on the kinematic diagrams of masters and slaves : "complete controllability" and "preservation of spatial configuration".

Complete controllability requires that, with a motionless hold, there is no spontaneous (i.e. other than by operator's will) movement of kinematic links caused by external forces (weight, hydrodynamics drag, etc.).

Spatial configuration preservation means that the kinematic diagrams of master and slave should guarantee against intensive movements of intermediate (between the attachment point and the grip) links under grip orientation change.

In choosing kinematic diagrams for masters and slaves, one has to define the number of degrees of freedom, types of kinematic pairs and their sequence. This task may be simplified by the consideration that kinematic diagrams of masters should be in some way related to those of the human arm. The kind of relation depends on the specific kinematic links of the arm (hand, forearm, shoulder) tracked by the master.

As the result, three group of kinematic diagrams may be identified (Fig. 1a, b, c).

If the operator's shoulder, forearm, and hand are all tracked, the master is attacked to the operator's shoulder girdle. With such a location of the master attachment point, the unambiguous correspondence between the positions of master and slave during operation will be violated because the slave attachement point is fixed with respect to the manipulator carrier, while that of the master moves together with the operator's shoulder. In most of the existing master-slave manipulator designs, the master attachment point is attempted to be fixed in space independent of the operator's body position. In this case, however, tracking of the operator's shoulder joint movements leads to an overcomplicated system having four kinematic links with twelve degrees of freedom (Fig. 1a). Practical application of this diagram is still restricted. Therefore the operator's forearm and hand (Fig. 1b), or the hand (Fig. 1c) are tracked.

In the case of operator's hand, tracking master-slave control is simpler because one can do with a lesser number of master degrees of freedom. The master need not be adjusted to individual features of the arm of a specific operator, and the arm may be disengaged from the master easier and faster. But in this case any point of the operating zone should be free of degrees of manoeuvrability otherwise the complete controllability condition will not hold. Analysis shows that if the rotation axes of the shoulder joints do not intersect the operating zones, masters and slaves tracking operator's hand can be constructed of kinematic chains having seven or eight degrees of freedom.

The kinematic diagrams of masters and slaves are required sometimes to be manoeuvrable, thus giving the operator the possibility of controlling movements in all degrees of freedom. In this case the hand and forearm of the master should track those of the human operator. In doing so, the number of master and slave degrees of freedom may be increased up to nine or ten (Fig. 1b).

The manipulators based on the above kinematic diagrams are general-purpose and perform all operations which are within the reach of the human operator. Reduction in the number of degrees of freedom restricts the technological possibilities of the manipulator which is no longer general-purpose and may be used only for special tasks.

To define dimensions of the master and slave kinematic links, one should know the operating zone served by the manipulator, or the zone of the operator if it is restricted by working conditions (this is characteristic, for instance of compartments of the deep-sea and space vehicles).

The notions of extremal and normal operating zones of the human operator are well known. They are based on his biomechanical, anthropometric, and physiological features [4]. Besides these main points one has to take into consideration many other ones, for instance, operator's safety, i.e. given distance between the master operating zone and the operator's body. This is particularly important for bilateral control systems where the master may move under the action of forces applied to the slave.

While controlling the manipulator slave, the operator's hand movement within the possible normal operating zone is restricted by the peculiarities of the slaves operating zone. In this connection, we are going to introduce the notion of "slave reduced operating zone" (decreased or increased) which is inscribed completely into the normal operator zone. Fig. 2 gives an example of the reduced operating zone. The master grip will move within the slaves reduced operating zone.

The ratio of linear dimensions of the slave reduced operating zone to the linear dimensions of the operating zone will be called "slave coefficient" (Km). Its value defines the relation between the dimensions of the master and the slave. The angles of rotation in master and slave joints should be the same. Convenience of operation depends very much on the correct choice of the value of "observed scale coefficient" (K'm) defined as the ratio of the slave grip movement, as it is seen by the operator in the screen (ℓ_e), to the linear movement of the operator's hand (ℓ_h). Coefficients K'_m = 0,6 - 1,0 have appeared to be the most convenient. Having defined the scale coefficients K_m and K'_m, one can formulate requirements for the visual feedback channel in terms of the maximally natural and convenient operator's operation, and define also dimensions of the master and slave 2, 5.

Requirements for the master-slave manipulator control systems also may be formulated with due regard to the features of the human operator included into the biotechnical complex "operator-manipulator".

Analysis of the human arm regulation has revealed that information from the receptors situated in the joints and other points arrives to the central nervous system computator discretely with spacing of 0,05 - 0,12 sec [7]. In this connection, the manipulator's control system should guarantee that the time of manipulator response to the controlling movements of the operator's arm does not exceed the latency period of the visual recptors (Fig. 3) [2]. But this criterion has turned out to be applicable mainly to time estimation for the slave. To solve the problem of synthesis, we had to carry out a series of experimental studies of the human operator working with various models of master-slave manipulators.

The following initial data were taken into account for construction of the manipulator model :

1. Scale coefficients K_m and K'_m.

2. Equivalent time constant T_e and oscillation coefficient of the model giving an approximate description of the manipulator dynamics.

3. Manner of control (master-slave or command) defining the kind of manipulator model input and controlling movements of the operator.

4. Communication channel delay which should be imitated in the manipulator model for the case of great distance between the operator and the place of operation.

5. Information display.

Three types of display were used in the experiments : stereoscopic TV set where operator sees directly the slave movements ; oscillograph screen with luminous point representing the slave grip ; and finally two-step copier showing in the oscillograph screen real-time movement of the two-link slave kinematic chain. For the model's subjective evaluation a three-point scale was used : manipulator dynamics does not interfere with the operator ; manipulator transient oscillations are appreciable ; and manipulator transient delays are appreciable. Fig. 4 shows experimentally found ranges of subjective estimates. Two experimentally found modes of operator work are also represented here by the areas of tracking and forced control [8].

In the tracking mode, the movements of the operator's arm and the slave are of similar character. With increase of the damping coefficient ξ (within definite limits), the operating conditions become better, and the operator's arm speeds and accelerations grow. Increase of the time constant of the manipulator dynamical model (T_e) results however, for the tracking mode, in reduction of the operator's arm speeds and accelerations.

With significant increase of the time constant T_e the operators passed to the second, forced control mode. Under this mode, the character of operator's arm movements differed very much from that of the slave's. The operators tried to compensate for deterioration in system dynamics by sharp increases of arm speeds and accelerations. Fig. 5 shows the speed of the operator hand (V_{op}) and the manipulator grip (V_h) versus equivalent time constant and damping coefficient [8].

Minima in $V_{op} = f (T_e , \xi)$ correspond to the passage of operators from the tracking mode to the forced control one. In this case the human operator, being an adaptive system, introduces various speed advances depending on the manipulator model time lag.

The mode of operator-manipulator interaction may be related to the performance index formulated above on the basis of the study of information processing in the central nervous system. Dependence of the slave grip's initial lag on an arm movements (with allowance for the visual threshold of movement perception) is shown in Fig. 6. Transition to the forced control mode begins at $\tau_b = \tau_{b1}$ and ends at $\tau_b = \tau_{b2}$. The range $\tau_{b1} \leqslant \tau_b \leqslant \tau_{b2}$ corresponds to the latency of visual receptors. This experimentally bears out that the dynamics of the manipulator slave control be estimated through the criteria of initial and final lag.

REFERENCES

[1] КУЛЕШОВ В.С., ЛАКОТА Н.А. Динамика систем управления манипуляторами, Изд-во "Энергия", Москва, 1971.

[2] ЛАКОТА Н.А., ЛОБАЧЕВ В.И. Некоторые принципы проектирования дистанционно-управляемых копирующих манипуляторов.Сб."Механика Машин", вып.27-28, изд-во "Наука",1971.

[3] КОБРИНСКИЙ А.Е., СТЕПАНЕНКО Ю.А. Некоторые проблемы теории манипуляторов. Сб."Механика Машин", вып.7-8, Изд-во "Наука",1967

[4] КОТИК М.А. Краткий курс инженерной психологии.Изд-во "Валгус",Таллин, 1971.

[5] АФОНИН В.Л, ЛАКОТА Н.А.,ЛОБАЧЕВ В.И.,МОИСЕЕНКОВ В.А. Исследование некоторых особенностей человека - оператора как элемента биотехнической системы "оператор-манипулятор",сб."Теория и принципы устройства манипуляторов", изд-во "Наука", Москва, 1973.

[6] ЧХАИДЗЕ Л.В. Координация произвольных движений человека в условиях космического полета.Изд-во "Наука",Москва, 1968.

[7] ГИДИКОВ А.А. Кибернетика и кортикальная регуляция движений.София, 1964.

[8] ЛАКОТА Н.А., ЛОБАЧЕВ В.И., МОИСЕЕНКОВ В.А. Экспериментальное исследование биотехнической системы "оператор-манипулятор".Изд-во ВУЗов "Машиностроение", 1971,№ 9.

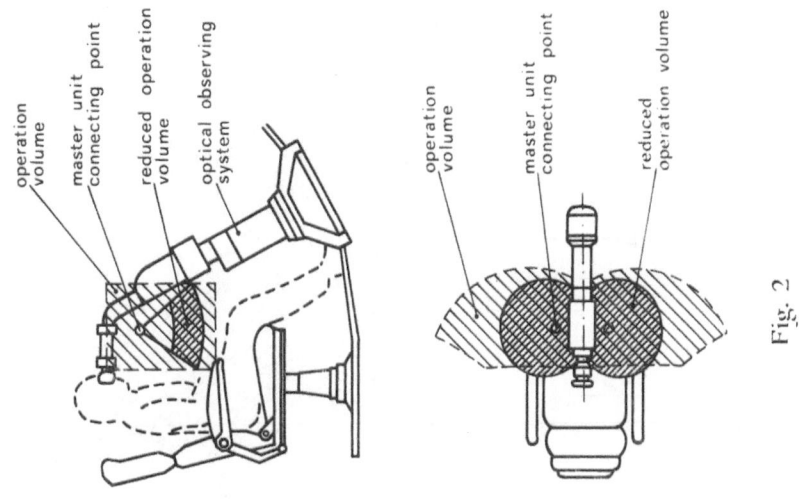

operation volume

master unit connecting point

reduced operation volume

optical observing system

operation volume

master unit connecting point

reduced operation volume

Fig. 2

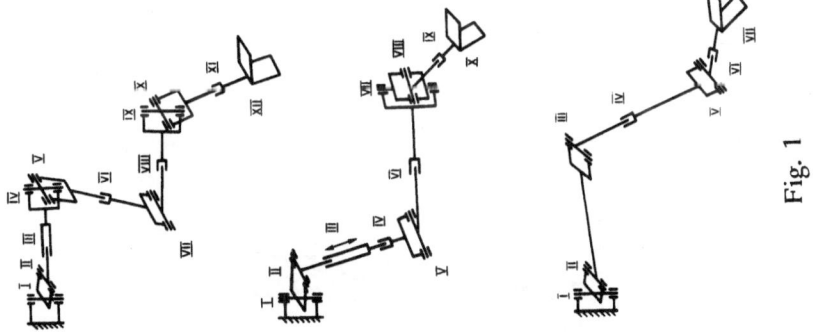

Fig. 1

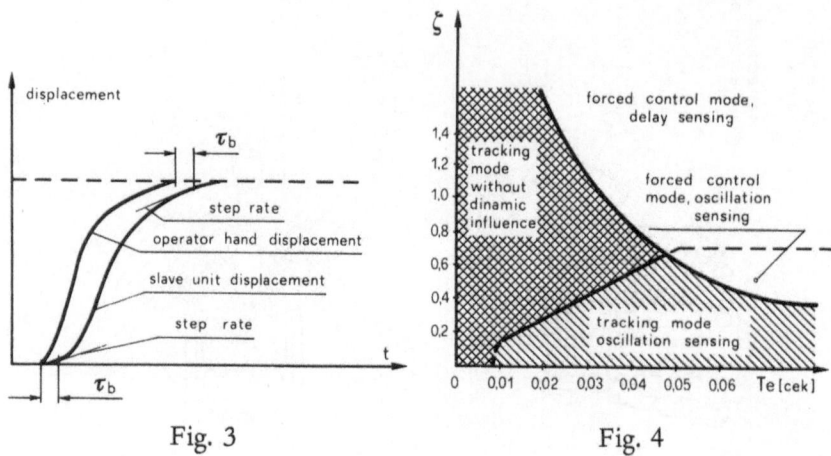

Fig. 3   Fig. 4

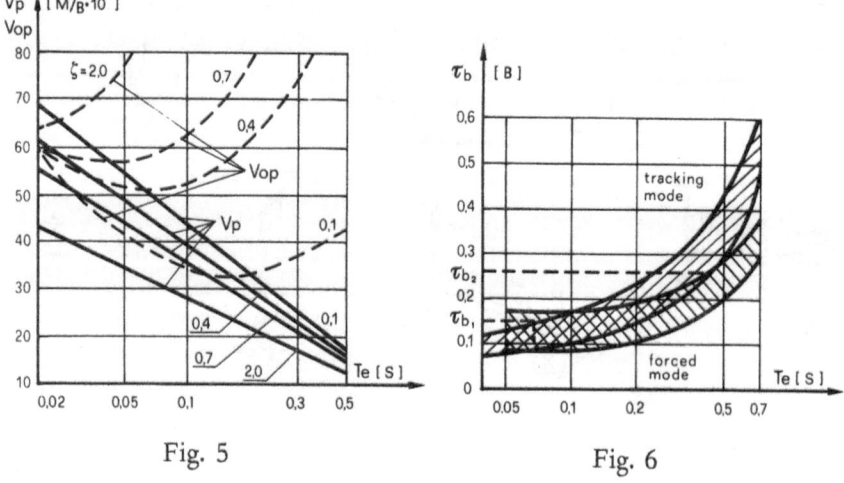

Fig. 5   Fig. 6

NORMALIZATION OF WALKING ON PROSTHESIS
WITH AN EXTERNAL POWER SOURCE

V. Yu. SHISHMAREV, Dr.,
The Central Research Institute for Prosthetics,
Moscow, USSR

I.S. MOREINIS, Professor
The Central Research Institute for Prosthetics,
Moscow, USSR

A.I. BOGOMOLOV, Dr.,
The Central Research Institute for Prosthetics,
Moscow, USSR

A.I. KOROTKOV, Eng.,
The Central Research Institute for Prosthetics,
Moscow, USSR

(*)

(*) All figures quoted in the text are at the end of the lecture.

The mathematical approach plays an increasing role in the designing of lower extremity prostheses. But the use of mathematical methods prosthetic design it becomes closely connected with the synthesis of an enormous amount of information given by biomechanics, physiology, the theory of automatic regualtion, information theory, and other sciences.

Mathematical methods in prosthetic design help to solve important practical problems, including rational mass distribution in prostheses and knee mechanism designed for above-knee prostheses which ensure the "natural" movement of the artificial leg in the swing phase.

The study of normal human gait and that of an amputee with regard to the lower extremities as four-link systems enabled us to solve a number of problems which can be termed the "prosthetic design scheme".

These studies made it possible approximately to determine human power losses during walking of the amputee and normal test subjects.

Further progress in artificial limb design is connected with the utilization of external power sources.

About 30 years ago N.A. Bernstein stressed the necessity of motor above-knee prosthesis construction (Fig. 1), but only few achievements in this field can be listed. There is now a number of experimental designs of prostheses for above-knee amputees where the power of the external source is used for locking the knee joint under the load.

The use of theoretical methods in automatic control systems for lower extremity prostheses is within the framework of task-setting.

Slow progress in this field of research is due to the fact the control disturbance exerted by the amputee himself should also be taken into consideration during the synthesis of automatic control system for artificial lower extremities.

This engenders the problem of giving a description of the complex "man-prosthesis" system. But so far there is no mathematical description of the man regulation system. The system controlling the upper extremity prostheses can be considered in some degree of approximation as a typical block diagram of automatic regulation (see fig. 2). But we cannot represent the automatic control system for an artificial lower extremity in a similar way. The movement of the upper extremity is weakly influenced by the movement of other limbs, but at the same time the movement of upper extremities or of the trunk and other leg can essentially influence the state of the leg bearing load and can upset the equilibrium and cause a fall.

The block-diagram of the system controlling the walking of an amputee is shown in Fig. 3, where the dotted rectangle represents the biomechanical linkage consisting of body segments connected with each other (Block T) and links of prosthesis (Block P). The upper circuit loop with feedback represents the model of motion regulation performed by the nervous system. The lower one represents the open automatic system controlling the artificial leg.

The analysis and synthesis of the system which controls the walking of an amputee call for the elaboration and analysis of a mathematical model of human gait which would make it possible to unite a large amount of information concerning conditions of the human propulsion system in a regulated functional complex.

In working out the mathematical model of the usual level human walking, we proceeded from the possibility of considering the human body as a biomechanical linkage moving under the action of natural and control forces. In this respect, account is taken of the mathematical model of walking with due regard to the movement only in the saggital plane as the most important factor in the analysis of the main objective laws. The movement of the upper extremities during walking is neglected. Segments of the human body were considered as absolutely rigid and joint connected. Muscular efforts were reduced to the muscular interjoint torques.

The human walking was thus regarded as a swinging fork biomechanical system consisting of nine links with an immovable point of contact between foot and the ground. Segments of the lower extremity, i.e. femur, shank, front and rear portion of the foot were represented as four links, but the trunk, head and upper extremities were treated as a single link.

The potentiometric goniometers, the force plates and time measuring devices were used to get the experimental data of walking.

The inertia characteristics of test subjects were determined according to their individual peculiarities and on the basis of experimental methods elaborated by the authors and the published data.

The biomechanical linkage system imitates the human body and changes during walking. In cases of a double support the additional forces of constraint are applied; the above linkage system changes its state from open to closed which causes jumps in the variation of kinetic energy.

The equations denoting the dynamics of the biomechanical linkage system in the stance phase after the corresponding transformation may be shown as follows:

$$R_z \sum_{i-k}^{n} \ell_i \sin q_i - R_x \sum_{i-k}^{n} \ell_i \cos q_i + g \sum_i^{n} \sin q_k \cdot$$

$$\cdot (\ell_i \sum_{j=i+1}^{n} m_j + m_i s_i) - \sum_{i-k}^{n} \sum_{r-1}^{n} \ell_i \cos\beta_{ir}(m_r s_r +$$

$$+ \ell_r \sum_{j-r+1}^{n} m_j) \ddot{q}_r \big|_{i \neq r} + \sum_{i=k}^{n} (J_i + \ell_i^2 \sum_{j=i+1}^{n} m_j) \ddot{q}_i +$$

$$+ \sum_{i=k}^{n} \sum_{r=1}^{n} \ell_i \sin\beta_{ir}(m_r s_r + \ell_r \sum_{j-r+1}^{n} m_j) \dot{q}_r^2 = M_k$$

$$k = 1, \ldots n,$$

where

R_z, R_x are the vertical and fore-and-aft components of ground-to-foot reaction;

m_i, ℓ_i, s_i are the mass, length and static radius of gyration for the i-link correspondingly;

I_i,   is the moment of inertia for the i-link about the i-joint;

M_i,   is the control force moments;

q_i,   is the generalized coordinates (interlink angles).

The control force moments and the power generated about the main joints of the lower extremity during normal walking were determined on the basis of the mathematical model and experimental data with the help of a digital computer.

The control power generated about metatarsal (N_1), ankle (N_2), knee (N_3) and hip (N_4) joints, respectively, during normal walking is given in Fig. 4.

It should be noted that considerable impulses of power occur in the stance phase when two legs are bearing the load.

The power of the control disturbance can be either of a positive sign or a negative sign.

In cases of flat-foot, the variations of control powers generated in the knee joint and in the ankle and metatarsal joints within 0.3-0.4 seconds are relatively small, but in the moments of heel strike and toe-off, there are considerable jump changes of power.

The study of dynamic and power characteristics of normal walking thus indicates that the process of human walking can be considered as an impulse system.

Consideration of the mathematical model of human walking has helped us to discover that external power sources are needed for the compensation of power lost by the above-knee or below-knee amputee. Though the power sources incorporated in the prostheses of the lower extremities should provide the prosthesis drive with impulses of power of about 1000 wt, their mean power capacity is relatively small.

The negative sign for the power indicates the process of power storing.

The absorber units incorporated in the prosthetic designs give an opportunity of secondary utilization of power spent during walking.

The mathematical model of human walking takes into account friction forces acting in the joints and ground, and to some extent the chosen approach to the control problem of the lower extremity prostheses supplied with external power sources.

It is supposed that the link deviations, in the biokinematic linkage, from the vertical line are of small value, this enables us to represent the equations describing the human walking as a linear nonhomogeneous system with automatic regulation

$$\dot{\vec{x}} = A\vec{x} + D\vec{m} \, ,$$

where
$\vec{x}$ is the vector of state of the human propulsion system, and is a column vector of angle deviations or of velocities,

A and D are the matrices of coefficients and control respectively,

$\vec{m}$ is the vector of control disturbance.

Naturally, such an approach is the first step in the determination of reactions of the human propulsion system to spontaneous input disturbance. The different components of second and higher orders have to be taken into account, but this is connected with the tackling of complicated theoretical and practical problems. Reducing the mathematical model to a linear nonhomogeneous system has at the same time, the advantage that the theory of such systems is sufficiently worked out.

The representing of walking control system as linear nonhomogeneous system with variable parameters makes it possible to find optimal solution of control problem. Having determined the control disturbances applied to human body segments during walking we have come to the conclusion that walking can be considered as an impulse system with characteristics similar to that of a relay system. Optimal control disturbance for a linear nonhomogeneous system which transforms the latter from one state into another within a short period of time can be also considered as an output signal of an ideal relay element.

The block-diagrams for N-dimensional and two-dimensional automatic control systems characterized by optimal early response is given in Fig. 5. Where S is the switching function depending upon auxiliary state vector P. The latter can be determined on the basis of a matching system.

When the Central Scientific Research Institute of Rehabilitation and Prosthetic Production carried out the development of the prosthesis for above-knee amputee supplied with a relay-control knee unit and external power source the main attention was paid to the stance phase. The relay control provides flexion in a prosthesis knee unit similar to that of a normal subject.

As the result of the mathematical approach to movement asymmetry of the normal and artificial extremities at stance phases, which was estimated according to the difference of the hip joint centres' trajectories of each extremity, the relation showing the dependance of the asymmetry on the angle of flexion of the prosthesis knee unit was found. At some optimum value of flexion angle (equal approximately to 0.2 radians), alongside with the increasing rythm of walking the coefficient of asymmetry in the movement of the hip joint centres was decreased in the vertical direction by 1 cm and in the direction of conveyance by 5 cm, which proves the effectiveness of prosthesis control at stance phase. The chart showing the normal angle of knee flexion is in Fig. 6, in Fig. 7 - that for the usual prostheses, and the chart for prostheses with knee control is in Fig. 8.

Analysis of the mathematical model has shown that power capacity of extremities with prostheses with rigid knee joints are much lower than in cases of flexible knee joint. Overloading of the sound leg is decreased by 7-10 per cent as compared to walking with traditional type prostheses, if we employ above-knee prostheses with controlled knee units using the condition of minimum movement asymmetry values for the artificial and natural extremities. This fact is of great importance and opens a wide field of research before us.

REFERENCES

[1] И.Ш.МОРЕЙНИС - Проблемы биомеханической асимметрии и ее коррекция
     при протезировании.Биофизика, 5,3,М.,1960.

[2] Р.Б.НАРОДИЦКАЯ - Биомеханический анализ механизмов протезов бедра
     и пути нормализации их кинематики.Канд.дисс.М.,1971.

[3] И.Ш.МОРЕЙНИС - Применемие методов механики для исследования принци-
     пиальных схем построения протезов бедра.Канд.дисс.М.,1962.

[4] Н.А.БЕРНШТЕЙН - К биодинамической теории построения протезов ниж-
     них конечностей.В сб."Труды МНИИП", I, М.,1948.

[5] Я.С.ЯКОБСОН, В.И.ДЕЛОВ, Е.П.ПОЛЯН, Ю.С.МЕЛЬНИКОВ - Макет протеза
     бедра с гидравлическим устройством, управляемым биотоками
     мышц культи.В сб."Протезирование и протезостроение", 16,М.,
     1965.

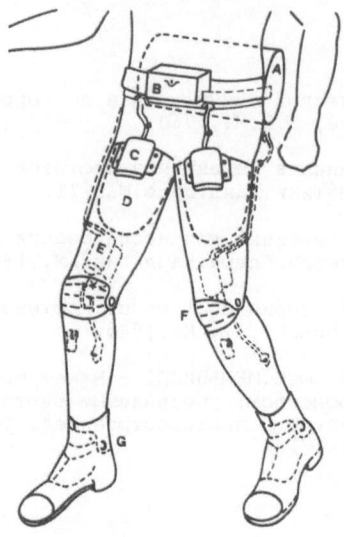

Fig. 1 Motor above-knee prosthesis
(according to N.A. Bernstein)

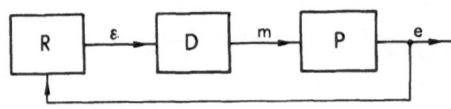

Fig. 2 Block-diagram of control system for upper extremity
prostheses

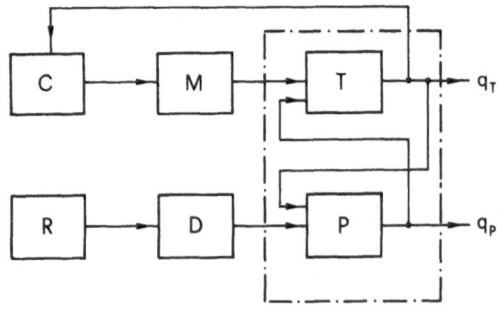

Fig. 3 Block-diagram of walking
control system

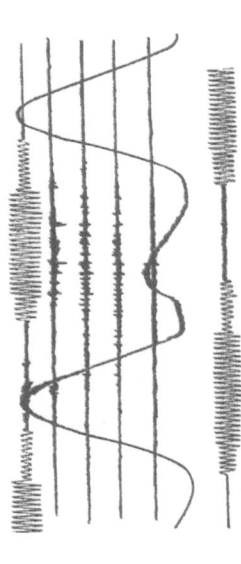

Fig. 6 Chart of knee joint angle during normal walking

Fig. 7 Chart of knee joint angle during walking on traditional type prosthesis.

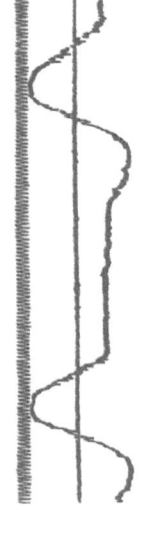

Fig. 8 Chart of knee joint angle during walking on prosthesis with control knee unit.

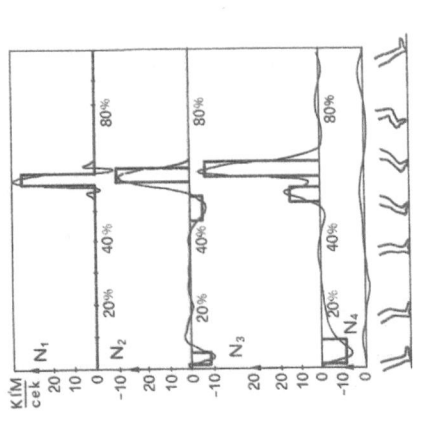

Fig. 4 Charts of control powers generated by the movements of muscle forces within one double stroke.

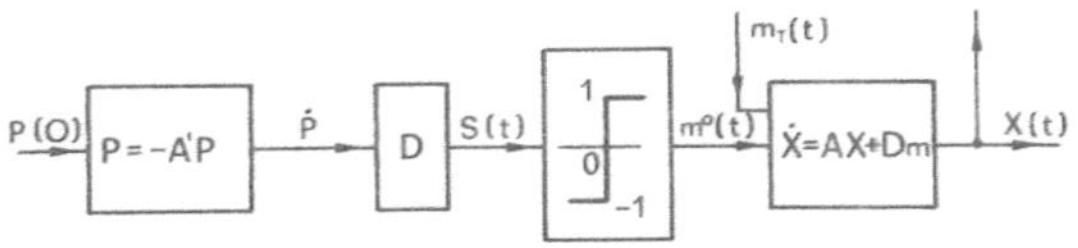

Fig. 5 Block-diagram of two dimensional system of automatic control.

LOGICAL CONTROL OF ROBOTS

Rajko TOMOVIĆ, professor
Faculty of Electrical Engineering,
Belgrade, Yugoslavia

(*)

Summary

Antropomorphic robots are treated as multivariable mechanical systems with large number of degrees of freedom. In order to make the control of antropomorphic robots feasible multilevel decision making, including the human operator, is assumed.

Such an approach involves a coordination controller which can be synthesized using logical description of locomotion. Methodology for logical description of locomotion based on automatic data processing is presented.

(*) All figures quoted in the text are at the end of the lecture.

1. Introduction.

Within the general class of robots, antropomorphic machines represent most complex multivariable mechanical systems. In order to control them multilevel hierarchical approach must be applied. General control philosophy of multivariable robots has been outlined elsewhere. [1] Basic control model proposed consists of three distinct levels: voluntary, logical (algebraic) and dynamic. Such an approach implies man-robot interaction so that human operator remains in charge of decision making. Another name used for such type of multilevel control of robots is semiautomatic control. Essential aspects of semiautomatic control for manipulators are described in the references [2,3].

In the design of the above multilevel control for robots important problem to solve is the synthesis of the controller at the logical (coordination) level. This paper deals exclusively with that task. As an example biped locomotion is taken. Methodology described can be easily extended to any other type of multilegged machine.

2. Logical description of the biped locomotion

The meaning of the logical control level can be best understood in the following way. As known, each human being, even each step of the same man, have different dynamic characteristics. Therefore, the full description and control of locomotion, as viewed from dynamic level, require, at least, as many different mathematical expressions as there are individuals.

On the other side, it is evident that for each type of the gait certain locomotion functions are invariant. If this is so, how can one derive the invariant locomotion patterns out of the dynamic behaviour which only can be recorded without affecting the integrity of the man or animal ? The derivation of the invariant locomotion patterns is essential step in reducing the complexity of the control system for walking machines. These invariants can be used for control purposes. Once that all details of the dynamic behaviour of individuals are filtered out, an "average" locomotion trajectory is reproduced with less requirements on the controller.

It will be shown that invariant locomotion patterns as detected from measurements, are equivalent to logical (boolean) expressions. Therefore the corresponding control level bears the same name. In other instances the name algebraic or algorithmic control is used.

2.1. Measurements

So far two approaches to derive logical descriptions for locomotion have been proposed. The first attempts were based on the finite state approximation of the dynamic behaviour of animal or human joints [4]. The validity of this hypothesis was checked experimentally. Namely, the algorithms obtained on the basis of the finite state approach can be imposed as control vector on the walking machine. The robot built in this way maintained repetitive locomotion activities. Experiments of this kind were performed on four-legged walking machines [5].

The approach presented in this paper is different. The basic idea is to use full dynamic records of biped locomotion. Out of these measurements one can derive the corresponding logical expressions [6]. Evidently, each type of the gait has its own invariants. In our case walking on level ground was considered. The person involved had full control of its motor activities. Locomotion of handicapped persons was also recorded.

Following measurements are used for the derivation of logical expressions: angle-time records for hip, knee, ankle joints of both legs (six records). In addition, the state of the support is needed. This information is obtained in the following way. Six switches distributed over the foot of each leg provide 6 + 6 on-off signals. They are used to describe in terms of logical expressions the state of the support of each leg. The distribution of the switches over the foot is shown in Fig. 1.

Full dynamic information needed for the logical description of the biped locomotion on level ground is displayed in Fig. 2.

2.2. Common logical support pattern

Logical description is first derived for the support function. Each step is divided into elementary phases. They correspond to different values of the following word: $x_1 x_2 ... x_{12}$, where x_i represents the state of the switch on the foot. The indices i = 1,2,...,6 refer to the left leg, while i = 7,8,...,12 to the right leg. Thus, each step is broken down into definite number of elementary phases.

In order to facilitate the analysis automatic processing of records was introduced. The computer prints out elementary phases of each step, counts the number of steps having the same number of phases and arranges the steps in the desired order.

Automatic processing is not so simple as may look at the first sight. Namely, the same switch is not always in the same state at each corresponding phase of the various steps.

Depending on individual features of each step, the same switch of the corresponding phase may be a number of times in state 1 or state 0. This situation can be considered as a new state and 3-valued logic must be introduced to analyze the step structure. In this way each word assumes again uniquely determined meaning so that they can be compared automatically to each other. When it comes to the final analysis, the letters standing for undefined states of the switches are omitted as redundant.

Once the steps are presented as ensembles of elementary phases, i.e., meanings of the word $x_1 x_2 ... x_{12}$, it is possible to proceed to the next stage. The final goal is to detect common pieces of information about the support function of the feet. Hence the name common logical support pattern CLSP. These patterns should be invariant for fairly large classes of persons. Just how wide coverage a CLSP has in the description of normal or handicapped locomotion habits of human population is difficult to establish at this stage of research. Great number of measurements must be undertaken at different research centers using the same methodology. These measurements based on automatic data processing are now going on in Yugoslavia at the Rehabilitaiton Center Ljubljana and Institute "Mihailo Pupin" Beograd.

Derivation of CLSP from the recorded data is not straightforward. It is essentially a pattern recognition problem. The simplest hypothesis to start with is that the steps having the smallest number of elementary phases are contained in all other steps. This proved not to be true so that other approaches are currently explored based on automatic pattern recognition procedures [8].

There are no reasons to restrict logical description of locomotion to support function. All other functions involved in biped locomotion such as standing, initial swing, joint dynamics can be taken into consideration. In the next stages of the project they all will be processed along the same lines.

2.3. Conclusion

Hypothesis that invariants of biped or other types of locomotion can be derived for wide classes of individuals is the starting point of this research. Methodology for logical description of the support function of human legs has been presented having in view automatic data processing. At this stage of the project it is essential to design software for automatic data processing and automatic pattern recognition so that large numbers of individual cases can be compared.

REFERENCES

[1]   Tomović R., Bellman R., "A Systems Approach to Muscle Control," Mathematical Biosciences, 8, 1970, pp. 265-277.

[2]   Tomović R., Gavrilović M., Marić M., "Computer Coordinated Remote Manipulation," Proc. IFIP Congress in Ljubljana, TA-7, 1971, pp. 1-5.

[3]   Marić M., "Semiautomatic Manipulation," Ph.D.thesis, Faculty of Electrical Engineering, Beograd, 1973.

[4]   Tomović R., McGhee R., "A Finite-State Approach to the Synthesis of Bioengineering Control Systems," IEEE Trans. Human Factors in Electronics, HFR-7, 1966, pp. 65-69.

[5]   Frank A., McGhee R., "Some Considerations to the Design of Autopilots for Legged Vehicles," Journal of Terramechanics, 6, 1969.

[6]   Tomović R., Zečević M., "Logical Description of Human Gait," Proc. IV International Conference, ETAN, Beograd, 1972 (in press).

[7]   Lazarević S., "Step Description by 3-valued logic," XX National Electronics Conference, Novi Sad, Yugoslavia (in press).

[8]   Turajlić S., Lazarević S., "Pattern Recognition of Support Invariants," XX National Electronics Conference, Novi Sad, Yugoslavia (in press).

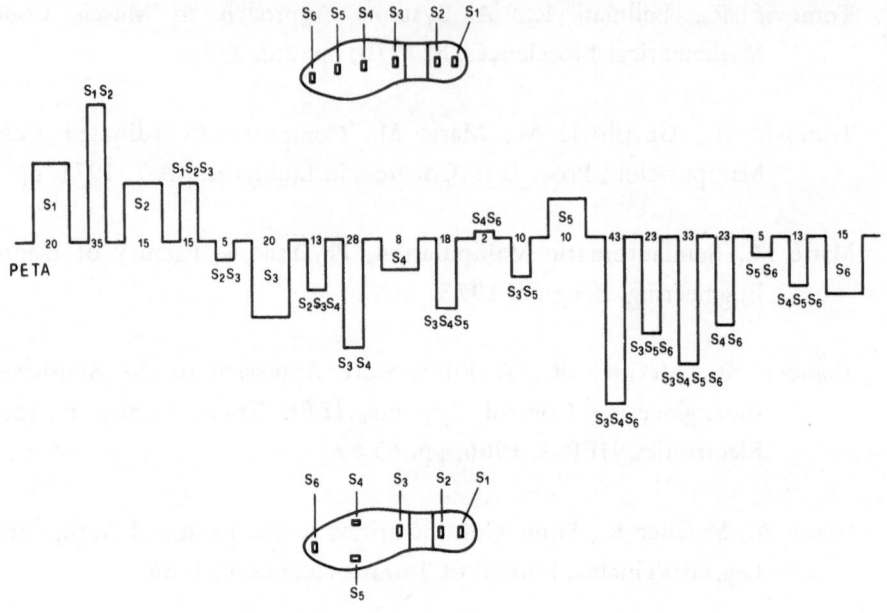

Fig. 1

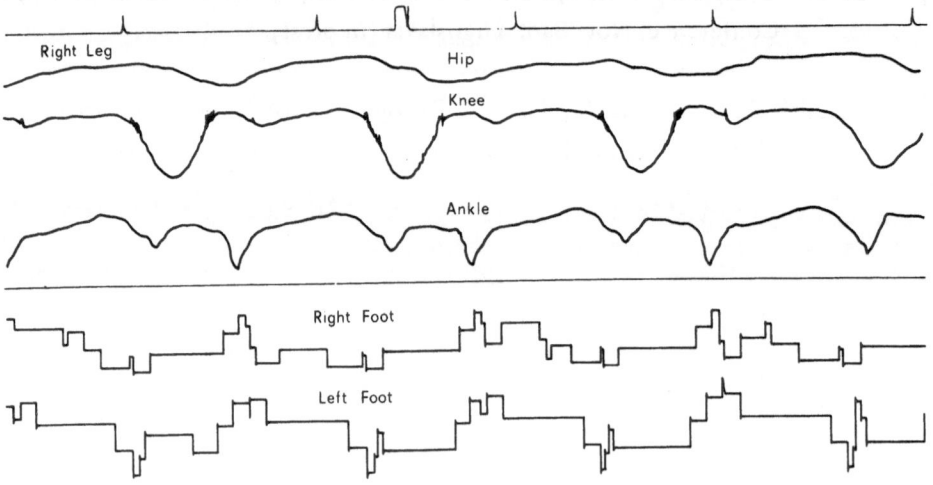

Fig. 2

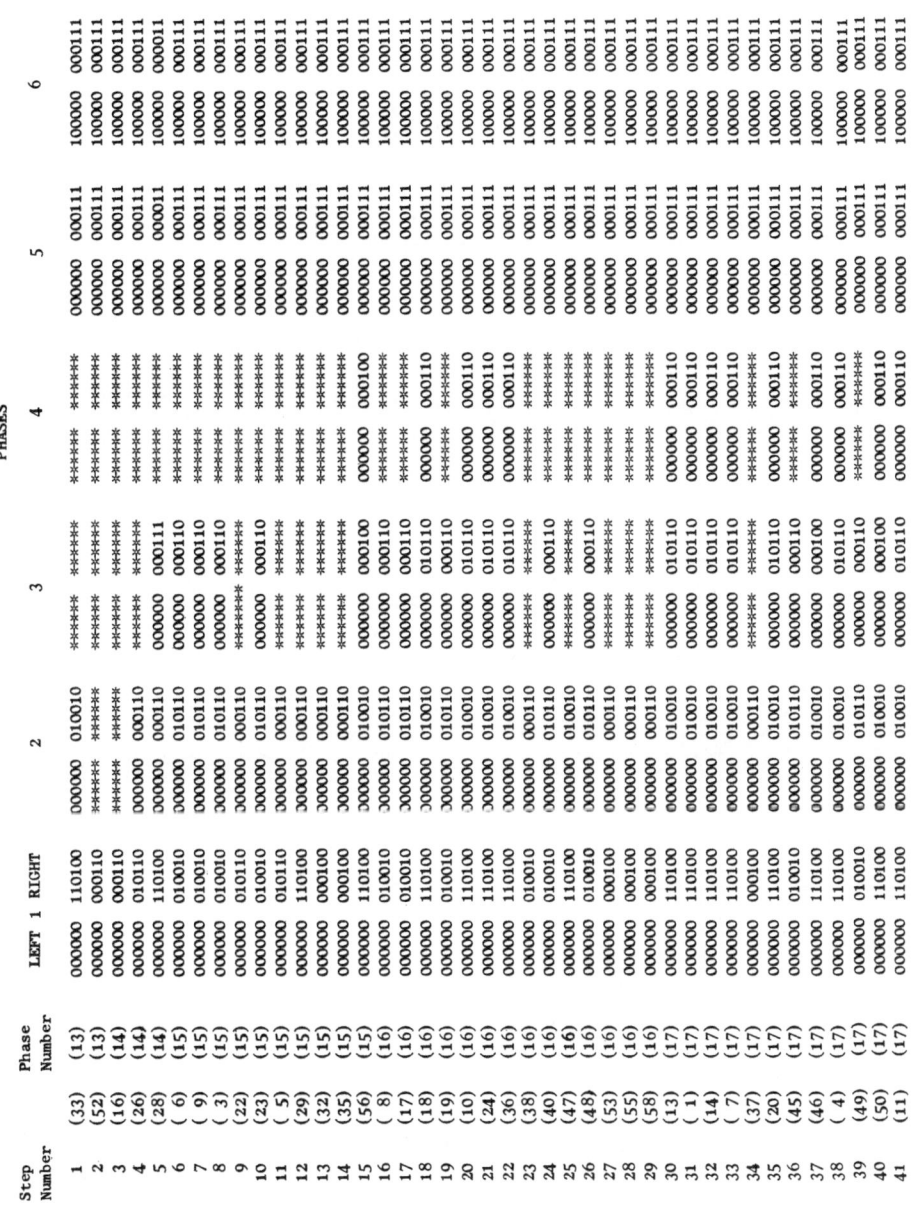

Fig. 3

MAN/MACHINE INTERACTIONS IN INTELLIGENT ROBOTIC SYSTEMS

Gershon WELTMAN and Amos FREEDY
Perceptronics, Inc., Encino,
California, USA

(*)

Summary

This paper deals with an experimental study of optimal interaction between human operators and intelligent robots in systems for decision making and control. The study uses an adaptive computer which is able to observe a control task, learn the required responses, and assume control responsibility. The initial findings show that the factors of greatest importance include how predictable the intelligent machine is, how many errors it makes initially, how much information the operator receives about machine decisions, and what his feelings are toward the machine.

(*) All figures quoted in the text are at the end of the lecture.

Introduction

In the coming years the human operator of complex technical systems will share more and more of his decision making and control responsibility with intelligent computers and robots. This apprach is already being developed for a broad range of civilian and military applications. These include remote and autonomous vehicular control, teleoperation and manipulation, various forms of command and control operations, pattern recognition, and others.

In a previous study an interactive adaptive system termed the Autonomous Control Subsystem (ACS) was used to share control of a remote manipulator with a human operator [1]. This experience revealed that in addition to problems of a purely technical nature, the ACS introduced a novel and unfamiliar dimension to the man/machine relationship: this was the ability of the operator to derive optimum help from a system which generates its own behavioral patterns.

We found that from the operator's standpoint, working with an intelligent machine is different from working with a deterministic one, even a deterministic one of great complexity. Barring malfunctions, deterministic decisions are generally accepted as true within their context. That is, we grasp readily that given the data at hand the machine was bound to have arrived at the particular conclusion. We may utilize or ignore this conclusion at our option, but we do not doubt it. Most previous examiantions of man/computer interactions have dealt with deterministic machines [2, 3, 4].

The intelligent computer or robot incorporates the idea of machine fallibility. In other words, it inherits some detriments of intellect along with the advantages. For example, its previous experience becomes a significant factor in its immediate response. Presented with apparently equivalent data the automaton can remain decisive, jump to conclusions, commit errors, learn slowly or rapidly at different times − perhaps even develop behavioral quirks. Although the process by which it arrives at decisions in some sense deterministic, it is often not amenable to rigorous analysis and certainly does not appear deterministic to the casual observer. Accordingly, the operator must add what he knows about the "mind of the machine" to what he knows about the problem at hand in order to maximize the chances of solving it.

Central to the entire idea of interaction with a learning system, then, is the operator's judgement of the help he is receiving from it [5,6,7]. Analysis indicates that the more accurately the operator can estimate the performance of the intelligent component, the more benefit he can derive from task-sharing with it. In the initial stages of exploration, then, an important point is to determine how the major system variables affect the operator's estimates of machine performance, what information he must have in order to optimize these estimates, and what is the best way to provide the information under operational conditions. Accordingly, the problem of operator cognizance of available machine aiding has provided a focus for the pilot study reported here, and will continue to provide the main impetus for future experimental work. Questions of optimal utilization and of specific means for informational transfer across the man/machine interface, however, will become relatively more important as research moves toward practical applications [8,9,10,11].

ACS System Concept

Figure 1 illustrates the basic concept of adaptive computer aiding by means of the Autonomous Control Subsystem. Man/Machine control with the ACS involves two main control loops: an external loop which incorporates the operator and his usual means of feedback, and an internal loop which contains only the Autonomous Control Subsystem. The major components of the man/machine interaction are defined as follows:

(a) **Performance Aiding** is provided by a computer placed parallel to the human operator in the man/machine control loop. The computer aids by making and displaying control decisions and by supplying autonomous control inputs to the machine system.

(b) **Adaptive Decision Making and Control** comes from a trainable "machine learning" computer program. Various sensors allow the program to observe operator performance and its results, and to optimize control decisions accordingly.

(c) **Decision Information** is presented to the operator continuously by the computer. The amount and type can vary with the application, and may include such factors as the degree of confidence in a computer decision, the planned action and the probable outcome, etc.

(d) **Allocation of Function** between the operator and the computer is made on the basis of the particular system and task, the immediate processing load, the type of decision information, etc.

The theoretical basis for the ACS is the Maximum Likelihood Decision Principle. Its structural organization is a conditional probability matrix relating future states of the man/machine system device to its past and present states [12].

Maximum likelihood was chosen for the ACS over various other possible classification systems because it has several significant advantages, including:

- Training is rapid and relatively simple
- Decision strategy can be changed while the system is active
- Classification categories are not restricted to disjoint sets

Initially, the ACS is untrained in the task at hand and acts mainly as a passive observer, using available sensors to record the operator's control actions and the resulting system outputs. Subsequently, the computer defines the likelihood of specific outputs occurring as a result of present and past system states, and constructs a conditional probability matrix accordingly. When probability for a certain action reaches a preset "Level of Confidence" (LOC), the ACS takes over system control autonomously. Depending on where the LOC is set, the ACS is a relatively conservative decision maker. Correct machine decisions reward the probability matrix while incorrect decisions punish it, so that the ACS moves continuously toward improved performance and adapts to changes in task requirements.

In time, for certain classes of task, the operator's role becomes that of initiator and inhibitor, providing occasional start and override commands. He is better able to provide goal direction, to undertake important secondary tasks, and to improve overall system performance. The power of the learning technique is that it bypassess much complex mathematical analysis and high-level communication: The operator shows the machine simply and directly how to optimize on his terms; the machine can then bring its unique resources to bear in order to continue the optimization process. The relatively simple aiding system described here is a precursor of much more complete and capable automatons.

Task Simulation

A generalized simulation was derived to permit subjects to interact with the ACS as they might in a broad variety of actual control tasks. The operator moved a light spot over the face of a large oscilloscope display using a two-dimensional, variable-rate joystick. Computer-generated boundaries defined sets of paths through a 9 × 9 matrix underlying the operating space. The task was to traverse the "safe" corridor as rapidly as possible while hitting the boundaries as little as possible. Figure 2 shows an operator at the control console.

The ACS program learned the path trajectories by monitoring the operator's successive attempts to run them. When the preset level of confidence (LOC) was reached, the ACS automatically took over control of the cursor and the operator's stick became inoperative. When confidence was lost, control was returned automatically to the operator. The operator was also given control if the computer made an error. Alternatively, the operator could override ACS control if he felt the computer was likely to make an error. A button on the stick provided this function. While the ACS was in control, it moved the dot at a speed typical of the average operator. As the ACS learned, it was in control for a larger and larger portion of the total path length.

Errors were defined as excursions from the path into a forbidden element. When this occurred (a) a red signal light was illuminated; and (b) all elements other than the one from which the excursion occurred became forbidden. Accordingly, after an error, the operator had to return to the path at the square where he left it in order to extinguish the error light. This avoided the operator "breaking through" to a new point on the path at the cost of only one error. Path-running simulated a variety of real-life control situations, including the supervision of semi-independent robot or manipulator devices.

Pilot Study

The purpose of the pilot study was to validate the new ACS program and task simulation under controlled situations, and to provide initial insight into the effect of some major system variables on the man/machine interaction. Eight subjects participated in the main study of ACS interaction. An additional two subjects provided a small control group, performing the task during two sessions apiece with no machine aiding whatsoever.

The experimental variables were:

(a) **Treshold LOC.** A "Low" value and a "High" value were used.

(b) **Task Difficulty.** An "Easy" set of two mazes and a "Hard" set of three mazes were established.

(c) **Feedback.** ACS takeover was indicated by a signal light, or no light was provided.

(d) **Background.** Male and female operators were selected from Technical or Non-Technical subgroups of the college student population.

Subjects were given a balanced set of treatments during four two-hour sessions. Each session included nine blocks of two trials each (easy task) or three trials each (hard task). A tenth block was run without ACS aiding. The focus was on man/machine interaction during the learning process. Accordingly, test sessions began with completely unlearned tasks and ended after learning had stabilized.

Experimental data reports were produced by the ACS program itself. For each path run, the data included the number of operator control decisions and his successes and errors in addition to the number of ACS decisions and their separation into success, error and override. Operator and ACS control times were recorded, as well as the maximum likelihood value associated with each decision. As a measure of subjective response, the operators were asked after each run to estimate on a scale of 0 to 100 the amount of aiding they received from the computer. They were questioned regarding their attitude toward the situation following each session and at the end of the experiment. They were given the Gottshaldt embedded figures test, in which "target" figures are hidden in larger designs, as a measure of perceptual orientation.

Experimental Results and Discussion

ACS Learning. Baseline learning performance for the ACS computer was obtained by running the paths repetitively without subject-induced error, so that ACS errors arose solely from the statistical properties of the sets and the inherent characteristics of the program.

Figure 3 presents the computer learning curves for the hard task; that for the easy task was similar. In both cases, the ACS assumed control more slowly with the higher LOC but made a fewer number of errors. With the low LOC, acquisition of control was more rapid, but the number of errors made initially was higher.

The errors made by the ACS as a percent of the moves it controlled, however, were essentially the same during each block for all task conditions. That is, the machine was initially wrong on about 65 per cent of its decisions; error rate dropped over three blocks to below 10 per cent, and remainded almost steady for the last five blocks. Thus the main difference between computer behavior in the low and high LOC cases was in the number of moves attempted and the number of errors resulting. The modes converged as learning progressed, so that by the end of nine blocks, performance was nearly the same for the two LOC conditions.

Transfer of Control. Figure 4 traces the transfer of control from operator to ACS over the course of nine blocks. Initially the operator controlled almost entirely, and somewhat over half of his responses resulted in correct moves.

The proportion of operator success did not change appreciably over the session. But as the machine learned the task, the amount of control required of the operator grew steadily less, so that finally he was initiating only about 30 per cent of the moves and was contributing usefully only about 20 per cent of the control decisions (his own correct moves plus corrections of machine errors).

Performance Aiding. An estimate of the overall aiding provided by the learning program in the current task can be obtained by comparing performance between the man/machine system and the two unaided operators. Table 1 summarizes mean data for the fourth and the ninth block of equivalent sessions.

Since the ACS was designed to move the cursor at about the same rate as the human operator, total task time was nearly the same at each stage for the aided hard task. Operator control time was markedly different for the two operating conditions, however. Even with the ACS only partially trained during the fourth block, it relieved the operator of 57 per cent of direct control responsibility in the easy case and 43 per cent in the hard case. Since the ACS learned paths faster than the operators, the aiding percentages for the ninth block were higher: 74 per cent and 63 per cent for the easy and hard cases, respectively. The man/machine system also performed better than the unaided operator with respect to total errors. At the fourth block the improvement in error rate was 74 per cent for the easy task and 45 per cent for the hard task. By the ninth block these had risen to 84 per cent and 64 per cent, respectively.

Estimates of Aiding Figure 5 and 6 compare mean operator estimates of aiding with mean aiding for the easy and hard tasks for the low and high LOC conditions. With one exception, the estimates followed the actual values quite closely over the course of the nine blocks. There was a slight tendency to underestimate machine contributions for the easy tasks, particularly during the period of rapid machine learning, but eventually the estimates caught up to reality.

The situation appeared fundamentally different for the hard task in the low LOC condition. Here the subjects underestimated seriously during learning, and continued to do so even after learning was complete. This case was characterized by a large number of ACS errors particularly during the early stages of learning, although the percent of machine errors was no different than usual. Apparently, the operators attached a high negative utility to the absolute occurrence of ACS errors, and "penalized" the computer accordingly in their aiding estimates.

Table 2 summarizes overall (RMS) difference scores for the aiding estimates made by the two subject groups. The Non-Technical group gave significantly poorer esitmates of aiding. Their difference scores were higher than the Technicals' in every case, differing by 62 per cent on the average.

In addition, the Non-Technical group showed a much higher variability in their estimates. For this group the aiding estimates appeared to be strongly affected by their own errors and control times. If the operator performed poorly in either accuracy or speed, he would seemingly punish the ACS through his rating. Conversely, a satisfying performance would be reflected in a high aiding estimate for that trial. For example, completion of the hard task under low LOC and no feedback light was perhaps the most difficult and frustrating condition – a combination of high task load and little knowledge of results. The Non-Technical difference score was substantially higher for this case; that for the Technical subjects was within the usual range.

The data for Table 3 suggest that the effect of the ACS takeover light differed for the two LOC conditions. Presence of the feedback signal seemingly improved aiding estimates in the high LOC case, but worsened estimates in the low LOC condition. This trend was seen in all four instances (easy and hard tasks) for the Technical group, and three out of four instances for the Non-Technicals. It is possible that a primary effect of the light was to call attention to the subsequent machine decision. In the low LOC case, there were more ACS decisions overall, and more ACS errors. Consequently, the frequent association of signal light with error light could have accentuated the observed tendency of the operator to lower the aiding estimate in proportion to ACS errors. The implication is that partial feedback data does not always sharpen the operator's view of the control situation. More complete information is most likely required.

Group Differences. Although separation of the subjects into Technical and Non-Technical groups was rather arbitrary, it did result in recognizable differences in subject characteristics and in their response to the system. With one exception, all Technical subjects scored higher on the Gottshaldt embedded figures test than did the Non-Technicals. The Technical group mean was 61 figures completed in 13 minutes (individual scores of 64, 63, 60, and 57) while the Non-Technical mean was 55 (scores of 63, 53, 53, 52). This difference was significant at the $P < 0.05$ level by the randomization test. [13]

According to previous investigators, then, one could assume that the Technical subjects tended more toward self-outwardness, interaction and active manipulation, while in general the Non-Technical favored self-inwardness and avoidance of interaction. [14,15] This seemed to coincide with a generally better ability of the Technicals to understand the control and information aspects of the man/machine interaction. They were more at ease initially, they used the override option more, and they estimated machine contributions more closely. Paradoxically, it was the Non-Technicals who exhibited the more positive attitudes toward system operation, suggesting that familiarity can breed contempt for machines as well as for men.

Individual Differences. Individual differences in attitude seemed of greater importance than group differences in determining the course of the man/machine interaction. Subjects reacted in many ways, some of them totally unforseen, to their encounter with a learning machine. The more interesting responses are described here in capsule form.

"A", a male Psychology major, and "M", a male journalism student elicted the best overall system performance (errors and time). They shared a number of characteristics — each had a high Gottshaldt score, a low error rate, and average performance. Their overrides were very effective, and apparently they contracted less on actually learning the maze than on becoming efficient supervisors. Nevertheless, "A" was the most competititve subject. He was frustrated with the system from the beginning. He thought the computer made errors on purpose, took advantage of good nature, and was a better learner. "A" gave a number of responses consistent with this outlook. He was emotional towards the display screen, laughed at computer mistakes, was constantly ready to override and enjoyed doing so. Also, he preferred the final non-aided runs. On the other hand, "M" showed none of these characteristics. He was emotionally positive, comfortable with the system and satisfied both with his work and with the computer's performance.

"S", an Engineering graduate, and "K", a Sociology sophomore, were a pair of subjects with better than average system performance, but with an overly high propensity for manual control. "K" was the more extreme example of manual emphasis. A pinball afficionado, he had the top scores in every operator performance category, but attained only near average system scores; due apparently to neglect of ACS supervision and interaciton. Both gave low subjective aiding estimates regardless of task condition.

"L" was a female music student with no technical experience and a low Gottshaldt score. She was not particularly adept in manual control, committed numerous errors and had longer than average operator control times. She did not interact well with the ACS, saying she felt threatened by the machine (which she kept referring to as "he"), and was unhappy at their partnership. She thought the machine was interfering with her ability to learn the problem at hand, and was visibly upset when the ACS committed errors which she had to undo. Her aiding estimates were quite low, but with greater familiarity her behavior became less extreme and in the final trials was similar to the more successful subjects.

"C", a female subject, an out-of-work elementary school teacher, had problems with the system and with her self-image as well. A particularly poor manual controller, she had the highest error rates and longest control times along with the lowest Gottshaldt score. "C" personified the computer completely, became lost and frustrated easily, and was competitive with the ACS in the initial sessions. She resolved her crisis in an unusual way — she became a secondary helper, physically and mentally to the computer. This was evident in her description of how she helped "him" learn the maze by awkwardly filling in the parts "he" hadn't learned, and in her complete aversion to override. It was also shown by her inability to estimate computer aiding, and by her tendency to overestimate when pressed, since she felt she was the one aiding the computer. Resolving the situation in this way, she became less frustrated and more relaxed. In the end, she referred to the computer as a teacher leading her through the mazes.

Conclusions

Preliminary analysis has verified that in intelligent man/machine systems it is necessary for the operator to estimate accurately the performance of the adaptive components so that the can utilize optimally its available help. The initial findings indicate that with some exception the average operator is quite good at judging how much aiding the machine is giving him at various points during learning. This is highly promising. Machine errors, however, seem to upset the operator's normally good judgment. When the machine makes a large number of errors, particularly in its early learning stages, the operator tends to underestimate its aid and to express dissatisfaction with its partnership. This happens even when it simultaneously makes a large number of correct decisions, and in fact actually raises the overall level of aiding. Auxiliary feedback on machine decisions may worsen this reaction by calling attention to the occurrence of incorrect machine actions.

In a sense, the intelligent machine is defined by its ability to make unpredictable mistakes. Thus the response of the operator to errors is bound to be important in practical applications, and could become a major problem if not handled correctly. Several approaches to a solution are worthwhile considering. One is to incorporate the operator's own utility structure in the computer's decision rules so that automatic decisions are assigned individual LOC's depending on time, place, risk, etc., of error occurrence. Another is to give the operator more complete information about the machine state at the time of its decision. Displaying such factors as the immediate maximum likelihood, the predicted action, the machine's current "box score", and others, could help to put the actual results in better perspective. Finally, providing a means for the computer to undo wherever possible the effects of its errors (i.e. return the system to its previous state or to a neutral state) would likely reduce the operator's negative utility for such miscues. It is planned to explore these techniques in future work.

It was anticipated that in this pilot study the subjective measures of interaction would prove as interesting as the objective measures, if not more so. This certainly proved to be true, particularly with regard to individual reactions to the adaptive aiding situation. People tend to anthropomorphize and personify computers in the normal case. The tendency is almost unavoidable in the case of an adaptive machine which is designed to imitate human patterns of decision making and control. Actually, some degree of personification may prove helpful to the operator's achievement of a reliable "feel" for the probable responses of the machine. At any rate, our operators saw the computer in rather clear cut terms – as partner of opponent, as benign threatening, as simple or nearly omnipotent. Almost always, a major component of their reaction was how they related to the idea of response uncertainty, that is, to the observation that machine decitions varied in predictablility.

It is felt that individual operator characteristics will continue to be important in more complex interactions with similar automatons; but it is difficult to say now which characteristics are indicators of successful performance. Our present division of subjects into Technical and Non-Technical was somewhat arbitrary, although the Gottshaldt test results indicated that they did differ measurably along one dimension at least. The group influence on system performance was quite interesting. Technicals seemed to achieve more and like it less. Non-Technicals exhibited highly diverse reactions, but these included some genuinely symbiotic as well as more enthusiastic relationships with the ACS. Much of the difference in response can probably be attributed to the greater familiarity of the Technical subjects with complex equipment of all kinds, including computers. Nevertheless, in some ways it was the Technicals' familiarity with the task demand which brought them into conflict and competition with the learning program. On the basis of present results, the ideal operator may well be one who combines moderate control skills with a proclivity for supervising the machine — in short, a control executive. If this proves to be true, we may have to select future operators of adaptively aided systems on criteria quite distinct form those now used to qualify operators of systems which demand a high level of ability, but of an unshared nature. Subsequent studies will examine other correlates of individual performance based on this hypothesis.

The idea that the successful operator performs as a good executive is just one of the ways in which concepts of inter-personal relationships attach themselves naturally to the current task simualtion and, in fact, to any considerations of human interaction with intelligent machines. The problem of assessing the computer's performance, of task-sharing with it, and of utilizing fully its autonomous characteristics are mirrored in numerous everyday management problems. The negative utility which operators hold for machine mistakes parallels that which supervisors may hold for the mistakes of individuals or of certain group members. This suggests that adaptive man/machine systems might prove useful as model of man/man systems. Certainly the present simulation meets many of the requirements for such a model, including well-defined and adjustable parameters as well as easily obtained measures of interaction. Precise control of relationships is not easy to come by in psycho-social experimentation — so that the benefits of a programmable intelligent partner or subordinate should not be dismissed lightly.

In summary, it is easy to think of many practical systems which combine a human operator with semi-independent, intelligent manipulators or other robot devices. Our findings to date suggest that the success of such a partnership will depend on human as well as technical factors; these include how predictable the intelligent machine is, how many errors are obseved initially, how much information the operator receives about machine decisions, and what his feelings are toward the machine. Our study will continue to examine these and other influences on the man/machine interaction.

The research planned for the coming year builds on the current base of experience in three major areas; these are: (1) provision of a more complete simulation by adding a secondary task and augmented feedback; (2) extension of the experimental investigation oto include factors of time sharing, risk and utility, and (3) integration of a mathematical model of system interactions with the experimental work. In addition, it is planned separately to expand the capability of the adaptive ACS program to perform decision making and control. Specifically, the expansion would increase the number and complexity of adaptive parameters which can be varied in a task situation and introduce a higher level of self-organization by allowing the machine to adapt its confidence level to the operator's apparent utility structure.

The overall objective is that through the experimental program, adaptive, computer-aided decition making and control will be brought to an advanced level in a generalized man/machine system using economical computer hardware. The programming techniques and communications methods used will be available to system designers, along with design criteria based on a seried of formal experimental studies. It is hoped that in addition to opening new avenues of research investigation the planned program will contribute a body of practical and immediately applicable data.

REFERENCES

[1] Freedy, A., Hull, F.C., Luccaccini, L.F., and Lyman, J., A Computer-Based Learning System for Remote Manipulator Control, IEEE Trans. on S.M.C., SMC-1, 356-363, 1971

[2] Licklider, J.R.C., Man-Computer Symbiosis, IRE Transactions on Human Factors in Electronics, HFE-1, 4-11, 1960.

[3] Miller, L.W., Kaplan, R.J., and Edwards, W., JUDGE: A Valve-Judgment-Based Tactical Command System, Organizational Behaviour and Human Performance, 2, 329-374, 1967.

[4] Nickerson, R.S., Man-Computer Interaction: A Challenge for Human Factors Research, Ergonomics, 12, 501-517, 1969.

[5] Carbonell, J.R., On Man-Computer Interaction: A Model and Some Related Issues, IEEE Transactions on Systems Science and Cybernetics, SSC-5, 16-26, 1969.

[6] Cohen, H.S. and Ferrell, W.R., Human Operator Decision-Making in Manual Control, IEEE Transactions on Man-Machine Systems, MMS-10, 41-47, 1969.

[7] Peterson, C.R., Judgments of Probability and Utility for Decision Making, Univ. of Michigan Engr. Psych. Lab. Report 037230-1-A, Sept., 1971.

[8] Licklider, J.R.C., Man-COmputer Partnership, International Science and Technology, 18-26, 1965.

[9] Mills, R.G., Man-Computer Interaction — Present and Future, IEEE International Convention Record, 196-198, 1966.

[10] Taylor, R.J., A Digital Interface for the Computer Control of a Remote Manipulator, NASA Cr-80843, 1966,

[11] Weisbrod, R.L., Human Information Processing and the Design of Computer Information Systems, M.S. Thesis, UCLA, 1972.

[12] Freedy, A., The Application of a Theoretical Learning Model to Remote Manipulator Control, Ph.D. Thesis, UCLA, 1969.

[13] Siegel, S., Non-Parametric Statistics for the Behavioral Sciences, New York: Mc Graw Hill, 1956.

[14] Blake, R. and Ramsey, G., Perception, New York: Ronald Press, 1951.

[15] Buros, O.K., ed., The Sixth Mental Measurements Yearbook, New Jersey: Gryphon Press, 1965.

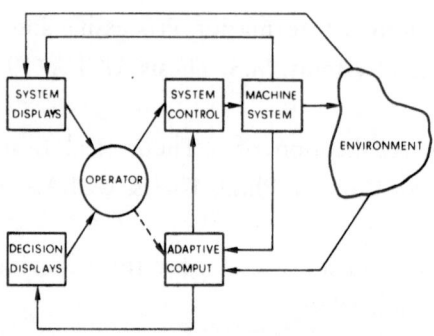

Fig. 1   Adaptive computer aiding in man/machine systems

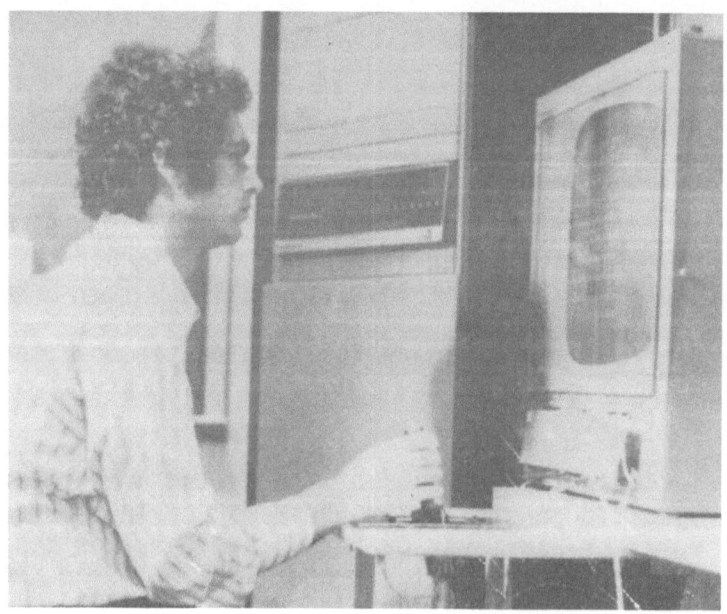

Fig. 2 The Task Simulation

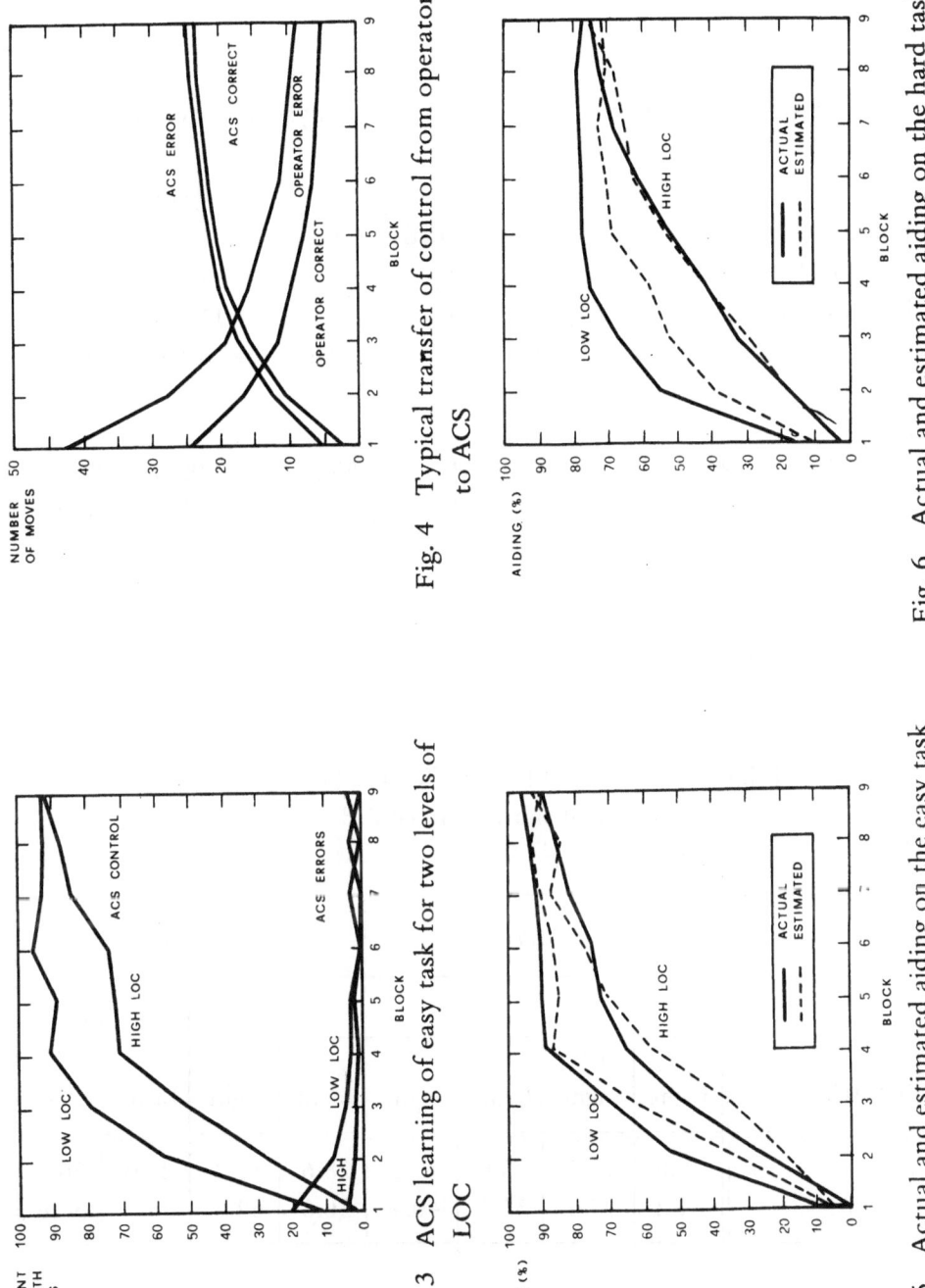

Fig. 3   ACS learning of easy task for two levels of LOC

Fig. 4   Typical transfer of control from operators to ACS

Fig. 5   Actual and estimated aiding on the easy task

Fig. 6   Actual and estimated aiding on the hard task

Table 1   A Comparison of Unaided and Aided System Performance

	Easy Task		Hard Task	
	Unaided	Aided	Unaided	Aided
4the Block				
Total Task Time (sec.)	69.5	68.7	115.0	95.0
Operator Control Time (sec.)	69.5	30.2	115.0	64.7
Total Errors	9.3	2.2	18.2	10.0
9th Block				
Total Task Time (sec.)	54.0	59.5	108.0	82.0
Operator Control Time (sec.)	54.0	14.5	108.0	40.0
Total Errors	6.5	1.0	17.8	6.3

Table 2 Summary of RMS Difference Scores Between
Actual and Estimated Aiding

TASK	EASY				HARD			
LOC	255		305		255		305	
FEEDBACK	Light	No Light	Light	No Light	Light	No Light	Light	No Light
Techs	25.0	15.7	16.6	19.5	28.6	19.2	21.0	26.6
Non-Techs.	34.2	17.6	35.8	40.4	30.6	62.8	25.2	34.6

CONTRIBUTION TO ANALYSE MANIPULATOR MORPHOLOGY COVERAGE AND DEXTERITY

Jean VERTUT et Al.
COMMISSARIAT A L'ENERGIE ATOMIQUE
SACLAY FRANCE

(*)

Summary

   Morphology of different manipulators and robots is analized and the limitations in orienting an object related to this morphology and to the position of the arm. From the extreme extension of the arm where the tong can only twist an object, to the center zone where the object can be oriented in maximum possibilities, several concentric zones are proposed. In the best zoning condition, out of special viewing problem, dexterity have been tested on different tasks to identify successive slowing factors due to the manipulator itself, the terminal device, the control mode etc... A proposed General Chart of dexterity is given for further investigation.

(*) All figures quoted in the text are at the end of the lecture.

Background

This work is part of a more general study in which we hope to reach sufficient knowledge of manipulator use to propose a basis for standardization. It is not necessary to elaborate on the many advantages of standardized models, such as substitution, simplified maintenance, etc ... The present paper covers a comparison of different models, or different concepts of manipulators in their geometry, evaluation of their space efficiency (the volume accessible compared to the volume effectively used to move the manipulator) and work efficiency, out of their dimension as well as their capacity.

Also, we hope soon to standardize the effective handling capacity with the corresponding static and dynamic safety factors and fatigue test specifications. This is extremely important, especially now when some unexperienced manufacturers could start to compete with the experienced makers and users of manipulators.

As in other fields, we are now discovering that many existing manipulators developed many years ago reach a very high level of development, but many important aspects of this technology have not been analyzed yet. In this particular case, the bilateral master slave, as originated before 1950 by Ray Goertz and his group, achieved, as a first attempt, a very close to perfect machine, as far as dexterity is concerned.

Now, when we expect new developments this field, an analysis is the only way to prepare a next step in the technology.

The previous work concerning dexterity was intended to verify the general possibilities of the manipulators, and the important recent work is related to new control modes.

The purpose of this contribution is to analyze different factors of the dexterity for a better understanding of the teleoperators an exceptionally close man-machine association.

Different motions (or degrees of freedom — DOF) of manipulators and corresponding morphology

As already established, manipulation needs at least six degrees of freedom. Positioning the object requires three translation coordinates. Then, orientation of the object requires three different axes of rotation. The seventh motion is operation of the tong or any terminal device.

Depending on the arm configuration, the positioning motion could be along a rectangular axis (such as with a bridge crane mounting). In this case, the elementary motions-lateral, backward-forward, and vertical-are called X, Y and Z motions, respectively.

Except this particular case, the arrangement of the positioning degrees of freedom (DOF) is related the polar coordinates, or more complicated.

In the next figures 1 and 2, we still designate motions in the same manner, around the normal position, as in the rectangular coordinates.

Fig. 1 is related to the usual master-slave manipulators used in nuclear technology [1-2]. Fig. 1 A shows the polar coordinates with the telescopic arrangement of most American designs. The Russian manipulators all use the arrangement shown in sketch B, with a sliding motion along the thru tube (cylindrical coordinates) and a rigid shoulder, with an articulated arm combination of the DOF are more complicated but the basic coordinates are still the same in the normal position, as shown on the other sketches in Fig. 1. Sketch C shows the human-like arrangement used in model BF 651 by Sorige France, sketch D shows the usual articulated arrangement used in the CRL model H, the French MA 11 [3] by La Calhene, the German Walishmiller A 200, and the American "mini-manip". Sketch E shows another arrangement (model BF by Sorige France) with the upper arm perpendicular to the thru-tube. Sketch F shows the arrangement of the Argonne electric master slaves [4], the CRL model M, the Italian "Mascot,, [5], and our recent French MA 22 [6].

The last arrangement shown (sketch G) is related to the Brookhaven electric master slave [7,8], and is typical of most motorized, open-loop manipulators as far as the arm itself is concerned.

Fig. 2 is related to the usual programmed manipulators now in growing use in industry (industrial robots). Sketch A illustrates the polar arrangement initiated by the Unimate, equivalent to fig. 1 A by permutation of Y and Z motions. Fig. 2 B shows the cylindrical coordinates originated by the AMF Versatran, also similar to fig. 1 B by permutating Y and Z motions.

Quite all machines in this category are falling into these two arrangements, few are equivalent to fig. 1 D, one or two are using the bridge crane arrangement in pure rectangular coordinates.

Concerning orientation motion, fig. 3 shows the well-known arrangement of the wrist, with azimuth, elevation, and twist motions. In most non nuclear manipulators the wrist is on the side of the forearm. There will soon be a need for a unified terminology for the different motions in all arrangements, as already proposed in the field of industrial robots and numerical control machines [9].

Manipulator Basic Installation principles into protective enclosures.

Mechanical master slave manipulators used in nuclear technology are working in "hot cells" through a thick protective wall [2]. So their mobility is limited to the arm only. To study the installation factors as well as coverage, the manipulator dimensions may be distinguished from the cell dimensions. For the manipulator, we have first the extended arm length A, shown in Fig. 4 A, which is determined by cell construction. The extension travel, Z, on telescopic arms and the proportioning between the two segments of the articulated arm are limited by arm length, A. The second basic dimension of the manipulator is the centerline distance B shown in Fig. 4 B. The length of the thru-tube includes some nonvariable parts, such as forks of the shoulder joints.

Concerning the mounting, we have the slave overhang C, into the enclosure, and the mounting height, D, measured over the working surface. The third dimension is the separation, E, between a pair of manipulators.

For a specific working surface such as shown in Fig. 4 A, with depth, P, and width, L, it is possible to install one pair of arms of a certain type with a given separation, E, and to calculate A so the extended tong can reach to the corners. To do that, we must also determine the slave overhang, C, and the bench mounting height, D. Dimension C is normally fixed at a maximum value for a given manipulator. Dimension D may be the minimum value which gives A, then Z, then the desired clearance when the arm is extended. This clearance is usually 600 mm (24 in.).

Outside the cell, as shown in Fig. 4 B for the American model 7 manipulator, it is possible to find dimension B by addition of the slave overhang, C, the shielding thickness (or window thickness if it is thicker), and the master overhang, which, in this particular case, is defined by the cell depth. For other models, the depth indexing (Y indexing) allows a constant master overhang which is fixed by cell construction.

Use of different types, such as model 8 and G, is illustrated in Fig. 5. In general, if two models can be used, the one with Y indexing motion needs a smaller window.

The cell height depends directly on the working bench diagonal, but is different for different models (such as model G compared with model 8). Also, other factors may be considered if the different models cover a different cell volume or cover the volume in a different way. This appears when a manipulator is used in a cell with a depth smaller than the maximum.

The situation is illustrated in Fig. 6 A with a telescopic manipulator from the maximum cell depth, to a classical cell, to a reduced cell depth, the same front-face arrangement is used. The only change is on the rear face. The opposite situation is shown in Fig. 6 B using an articulated arm (model H). Coverage is such that the cell depth reduction is better to be operated from the front face. This allows coverage of a shallow cell, with less overhang, using the same arm and covering the maximum depth with less shielding thickness. All these considerations are related to a good understanding of the morphology of the arm and to the resulting coverage structure.

Basic installation principles of manipulators in general

In most cases, and especially with industrial robots, the manipulator body is fixed relative to the working zone. This case is the same as described before. If the manipulator body can move, as well as if two arms are fitted on a mobile device we can get a larger coverage and installation principles are different as mobility of the arm and mobility of the whole teleoperator (*) are giving redundancy. It must be pointed out that the lost volume needed to move a big arm is often smaller than the volume to move the whole mobile unit with a smaller arm.

Problems related with limited available volume are as well typical of nuclear and industrial manipulators, but without sense for most undersea or space teleoperators.

Principle of coverage analysis

Coverage analysis [2-11] is based on the limitations of the orientation motions (azimuth, elevation, and twist) depending on the terminal device position.

The effective maximum volume covered at the tong tips is called zone 1. At this limit, when the arm is fully extended, the only possible orientation motion is the twist (also confused with azimuth).

That zone may extend outside acceptable limits (walls of a room, safety zone in a factory etc...). Then zone 2 is defined as the part of zone 1 inside these

---

(*) Teleoperator remote mobile dextrous machine as manipulators on vehicle [10] .

existing limits.

As we retract the arm from the limit of zone 1, the possible orientations of the object in the tongs (at center of the fingers) increase. When this gives a solid angle of one-quarter of a sphere as shown in Fig. 7 A, this will be the definition of zone 3. Of course this is more precisely defined as that quarter of a sphere having two plane faces vertical and horizontal. The horizontal bisector direction of this quarter sphere may be oriented in any direction. In summary, this means the minimal orientation in zone 3 is to pivot the object 180° in a horizontal plane and from horizontal to vertical down (pan and tilt).

An improvement on this is the freedom to tilt in the vertical direction which gives half a sphere as shown in Fig. 7 B, while we retain the main horizontal direction. This improved zone 3 is defined as zone 3'.

In zone 4 (Fig. 7 C), we can orient an object 360° in a horizontal plane and extend this plane vertically downward. This is equivalent to zone 3' but the main direction is vertical down. In this zone, the stop on the azimuth motion is still limiting. This zone allows all the motions of the manipulator on a bench.

Full-sphere mobility is rarely obtained on regular manipulators. We could call this zone 4'. It may be built into future manipulators. It also appears possible with an arm on a mobile positioner (as with the servo manipulators) by moving the full unit around the object.

The selection of the vertical and horizontal planes to limit the quarter sphere mobility is related to work on a table. These basic planes could be generalized in relation with important directions on the work to be performed (space, undersea).

In the particular case of mechanical manipulators fitted into a classical hot cell with a window, as shown in Fig. 8 in lateral view, we have some limitations due to both the viewing and the master arm. For example, some places could be accessible to the tong tip if the master handle would not touch the cold side of the window (this is minimized by use of the Y indexing motion). Another lost area exists when it is impossible to simultaneously view through the window and reach a remote lateral location with the tong. So we call zone 5 the in-cell zone where it is possible to both reach and see directly through the window. Of course the latera limit of zone 5 depend very much on the window width and X indexing motion. Figure 9 shows zone 5 as visible in a vertical projection. Zone 5 depends on the teleoperator's relative eye-hand positions. This is different when working with two hands in front in the normal working position, and when working with one hand in reversed position with the tong tip looking backward, as shown in Fig. 8, bottom sketch.

Figure 8 is a side view of these successive zones, and Table I summarizes definitions and limitations of the zones.

Compared zoning of different manipulators

Zoning charts have been traced for all existing models of mechanical master-slave manipulators and this analysis is ready to be proposed as standard. Of course each obstacle gives subsequent zone limits due to each segment motion limitation. These "shadows" of an existing obstacle to an arm is the reason to need geometrical redundancies as the human arm does have many. One of the easy redundancy is to fit the arm on a mobile device, as the human arm on the human body. In fact this work is just beginning in an unexplored field where geometry, ergonomy, and other human factors are critically mixed. Such analysis will be very useful in present and future applications of teleoperators, especially undersea and in space. We must also emphasize that this is very closely related to the dexterity study.

Dexterity factor – Efficiency factor – Time efficiency factor.

In this paper, "dexterity" of a manipulator is used as it compares to a human hand.

Flatau [12] proposed a dexterity quotient which is always less than one, but data in this work are given as a factor between time needed to perform a certain task using a manipulator, and the shorter time to do it by hand. I suggest calling this a "time efficiency" factor [13]. Of course this factor is always larger than one-only very new systems like multimode controlled manipulators [14] could approach one and may even improve the man capacity.

This time efficiency factor results from different sub-factors related to the operator, the task, the viewing, the manipulator, and the terminal device. An objective is to explore separation of these sub-factors.

Two types of tasks have been tested. In type one grasping, load displacement, and assembling operations are carried out in the same manner by hand and by manipulator. The second series (turning valves, plugging electric cables, etc...) are not carried out with the same procedure by hand and with manipulators. For these tasks, the total time efficiency factor will have a strategy slowing factor.

In all cases, these tasks have been carried on in the best viewing conditions, and in the best location of the manipulator's coverage without any limitation due to the geometry of the arm.

Study of remote viewing as a slowing factor is planned and potential improvement by a more sophisticated terminal device will be studied later.

Time efficiency experiment procedure

Four different tasks have been defined and tested with different manipulators, and with two types of handles for some.

In the first time study we compared the time to do a task by hand and by manipulator. The data are more a relative time efficiency for a task than a relationship between manipulators.

Recently we installed all models close together and compared them on the same task, with two different operators. This last work has been carried on with mechanical master slave only, as the first run was operated also with open loop manipulators. Data given in the present paper have been linked to the two series of tests from one common test, but evaluation is not yet definitive.

Fig. 10 illustrates the general procedure of this experiment. Two switches are installed on the bench to start and stop the task time record. This typical task is repeated 25 to 30 times for at least 15 min. Each task time is recorded as a peak on the recorder and total time as paper displacement (see fig. 11). Fig. 12 illustrates the hand type operation test with two different tasks.

Task 1 – "Pick and place" on Pins (fig. 12 A). This test enables to analyze picking time, fitting on pin time and average moving velocity when empty and with the two loads. To compute these different sub-tasks, we recorded seven different jobs with the constant distance of 0,4 m (16 in.) between pins. Each of the seven tasks was made 35 times. The same operator tested all models and then repeated the first one to check.

Task 2 – Assembling Test. As shown in Fig. 12 B this test consists in fitting a 1/2 in. peg into two holed, piled blocks and locking the whole thing with a pin. This second test needs more force coordination, as the peg is rather tight and rigid (the vertical pin in task 1 is very elastic). This test has been recorded in total time and separately for assembling and disassembling. Fig. 13 illustrates the unlike hand type operation test, two typical tasks using twisting motion have been tested.

Task 3 – Turnig Valve Test.

Task 4 – Electric Plug Test. Both of these tasks are easily made by hand with three fingers using at least 9 degrees-of-freedom and force feedback. With a poor two-parallel-jaws tong, this task needs a different strategy, with successive grasping and wrist twist. Parallel grasping of a cylinder is difficult with flat jaws, so this specific task adds different difficulties lack of force feedback, poor grasping etc. This has also been tested on open-loop manipulators with continuous rotation of the tong.

Slowing effect in hand like operations
Operator Training Factor:

Tests have been recorded for an unexperienced and a well-trained operator. The following remarks can be made. The average time for the new operator was 1.4 times that of the trained one, and his standard deviation σ) was 1.3 times larger. After one month on this specific work, this operator improved by 1.2, but not to the same extent on all master-slave manipulators. At the end, he was still 1.15 slower than the trained operator. The factor was the same by hand or by remote operations. So, the time efficiency factor data are the same for both operators, but they still have a personal time difference. This point will be further explored.

Operator Fatigue Factor:

Up to now we did not point out a clear record of fatigue, except from the handle point of view. On Task 1, the grasping of 1 kg (2 lb) with one specific handle is possible only a few times with one hand, so the operator does it with two hands. Another handle was used with one hand. Exploring the fatigue factor will also be the subject of further work.

Operating Viewing Factor:

This will also be studied later. At present, we try to avoid viewing problems by working without windows. We hope that distance differences, depending on models, did not interfere in this preliminary data.

Bilateral Master-Slave Slowing Effect:

Data for the pick-and-place task and assembling task for different models are shown in Table II. The different models [2] are concentrated in figure 14. On the first line of the table models are in the order of their time efficiency factor for task one. This order correlates with the inertia of these master-slave arms. Handle effect also appears, introducing a 10 per cent improvement to different models. The picking time and pin fitting time also vary with load and consequently with the type of handle. This handle factor must be explored further. In the present stage, it is related to fatigue and one-or-two-hand operation.

The average velocity evidently depends on inertia and blacklash. In this particular case, blacklash was a problem with our old model G, and limits acceleration, requiring slower movements for precision positioning.

Concerning MA 11 (model H), the inertia in the lateral displacement (X motion), is higher than in Y, and this X inertia is even lower than on the model G. We intend to make tests using larger distances, and to test the electronic MA 22 electric servo master-slave which allows also to consider the effect of different ratios, inertias etc...

Open-Loop Control Slowing Effect:

Test data are limited comparing this type to master-slaves. Table II includes our available data on tasks 1 and 2.

The Syntelmann, prototype of a German anthropomorphic manipulator with very little force feed-back, a PAR model 3500 (rate controlled), and the ACB (Ateliers et Chantiers de Bretagne, France) on/off control manipulator were tested. The tests show a minimum factor, produced by lack of force feedback, of 5 to 7 between the force-reflecting master slave and the unilateral arm. Since the normal speed of this arm is different from the others, this is an approximation. The MA 22 servo master slave illustrated fig. 15 will be used to further study this particular point which is extremely important for further developments.

Slowing, resulting from lack of force feedback, is also emphasized on the assembling task. This might be due to poor precision in position control.

Strategy slowing effect (unlike hand type task)

Our tests on unlike-hand type operations are summarized in Table III. This is made from first test runs which are approximately compared due to the long time elapse between testing of the different models.

The relative standard deviation, σ, is much higher than in task 1 by hand or manipulator.

Furthermore, σ is larger on the electric plug test than for the valve test because of poor ability to grasp the cylindrical plug. These figures will be improved by further work. At any rate, a 2-to-3 ratio is found between the hand-like operations (tasks 1 and 2) and non-hand-like operations (tasks 3 and 4). This is a first approach to the strategy factor.

General dexterity chart of manipulators

As an attempt to summarize existing and possible data on dexterity, table IV is a proposal for further investigation. This table is intending to fill the gap between previous [12-15.16.17], recent [14-18] and future work. This dexterity analysis will require more extensive test program.

Conclusion

The need for standardizing morphology analysis, as well as geometrical coverage and dexterity testing is to be emphasized when these three subjects have to take profit of extensive comparisons and international exchange.

Acknowledgements

This work have been greatly helped by discussions with Carl FLATAU, Ray GOERTZ, and JL NEVINS. The author also would like to thank in particular A. DUCONDI, J.C. GERMOND, J.P. GUILBAUD, L. PAPOT, D? ROSSIGNOL, R. SERAN, A. TIRET for their effective cooperation.

REFERENCES

[1] J. VERTUT et al —"Manual on Safety Aspect of the Design and Equipment of Hot Laboratories„– I.A.E.A. Safety serie n° 30

[2] J. VERTUT et al — Catalogue de Centres d'Etudes Nucleaires, "Protection-Manipulation" Chapter III, Section 2.

[3] J. VERTUT et al."Through the wall Master Slave Manipulator with Indexing„ — Proc. 12th Conference Remote System Technology 67 — ANS (1964)

[4] R.C. GOERTZ et al —"ANL Mark E 4 A Electric Master Slave Manipulator„– Proc. 14th Conference Remote System Technology — 115 ANS (1966)

[5] S. BARABASCHI et al"An Electronically controlled Servo manipulator„Proc. 9th Conference on Hot Labs and Equipment 143 ANS (1961)

[5 bis] L. GALBIATI et al"A Compact Flexible Servo System for Master Slave Electric Manipulator„ — Proc. 12th Conference Remote System Technology 73 ANS (1964)

[6] J. VERTUT — C. FLATAU et al —"MA 22 Compact Bilateral Servo Master Slave Manipulator„– Proc. 20th Conference Remote System Technology 296 — ANS (1972)

[7] C.R. FLATAU —"Compact Servo Master-Slave Manipulator with Optimized Communication Links„– Proc. 17th Conference Remote System Technology 154 ANS (1969)

[8] C.R. FLATAU —"Synthesis and Scaling of Advanced Manipulator Systems" (elsewhere in this Proceedings)

[9]   J. KOENIG and B. DIETRICH, "The Robomation-Innovation of Automation" Koenig Automation 168 170 Rixheim France p. 121

[10]   Advancements in Teleoperator Systems – NASA SP 5081

[11]   J. VERTUT "Analytical Zoning of Manipulator Coverage" Proc. 21st Conf. Remote System Technology, ANS (1973) to be published

[12]   C.R. FLATAU et al."Some Preliminary Correlations between Control Modes of Manipulator and their Performance Indices."

[13]   J. VERTUT "Contribution to Define A Dexterity Factor of Manipulators" Proc. 21st Conf. Remote Syst. Technol. ANS (1973) to be published

[14]   J.L. NEVINS "Teleoperator Technology-Past. Present and Future,, E. 2640 MIT Draper Laboratory (February 1972)

[15]   R.C. GOERTZ, Proc. 1964 Symp. Remotely Operated Special Equipment, pp. 27-69, USAEC Conf. 640508 (May 1964)

[16]   R.C. GOERTZ Report on ANL activity, Proc. 1964 Symp. Remotely Operated Special Equipment, pp. 41-47, USAEC Conf. 641120 (November 1964)

[17]   W.N. KARNA "Human Factor in Remote Handling" – Proc. 1964 Symp. Remotely Operated Special Equipment

[18]   J.L. NEVINS "The Multimoded Remote Manipulator System" E 2720, MIT Draper Laboratory (1972)

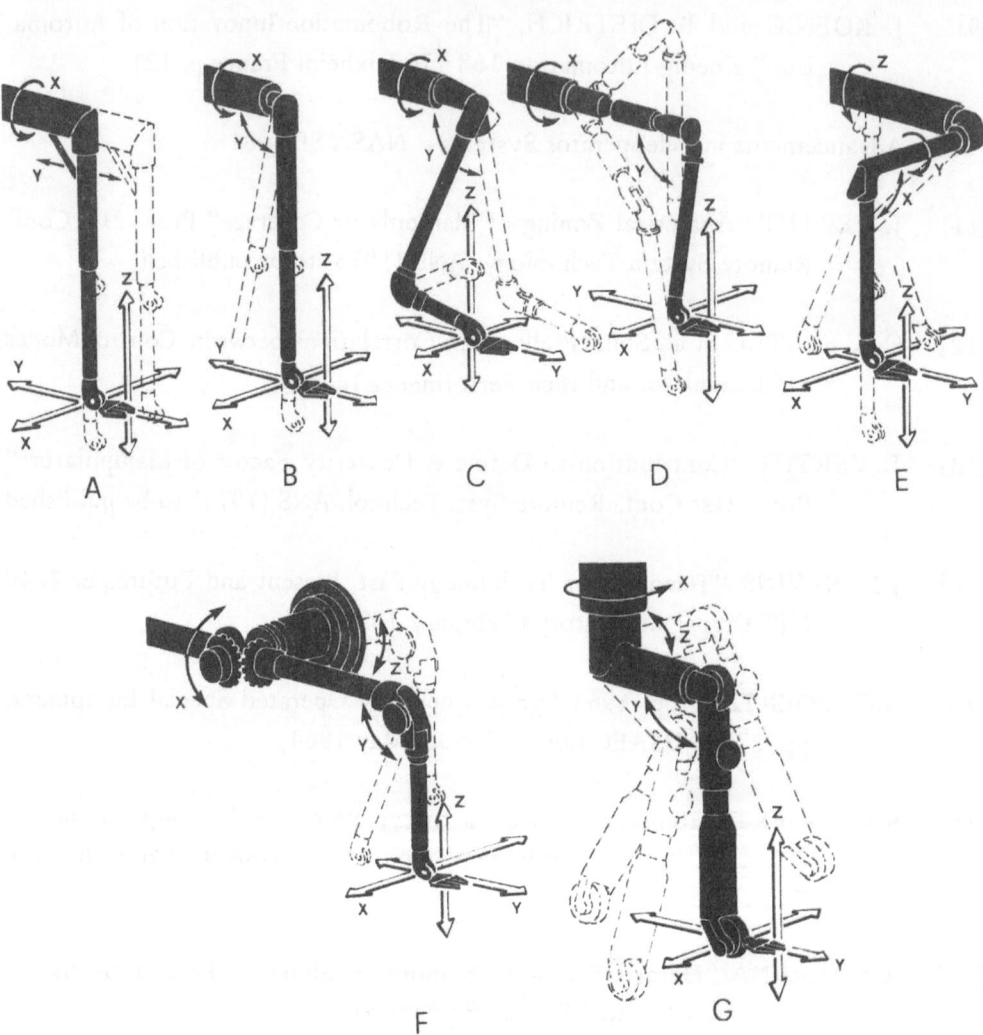

Fig. 1 Positioning movements on nuclear manipulators.

A. Classical Telescopic Manipulator (USA)
B. Rigid-Shoulder Russian Models
C. Human-Like Articulated Arm
D. Usual Articulated Arrangement
E. Side Arm Arrangement
F. Classical Electric Master-Slave Arrangement
G. Classical Bridge Crane-Mounted Power Arm - Brrokhaven manipulator

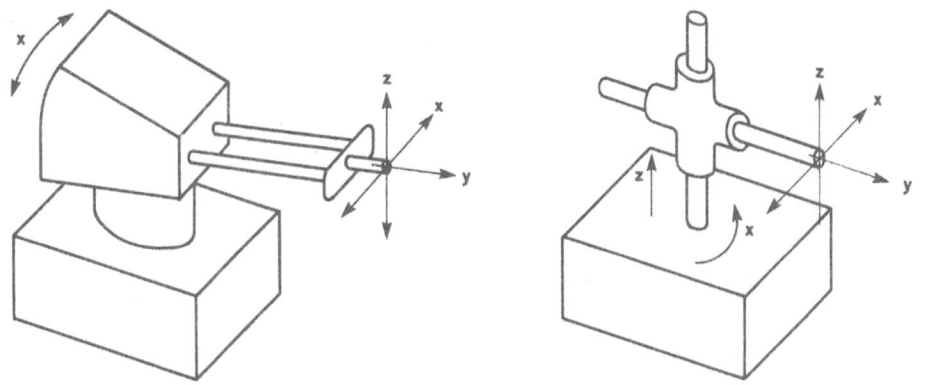

Fig. 2 Positioning movements on industrial manipulators.

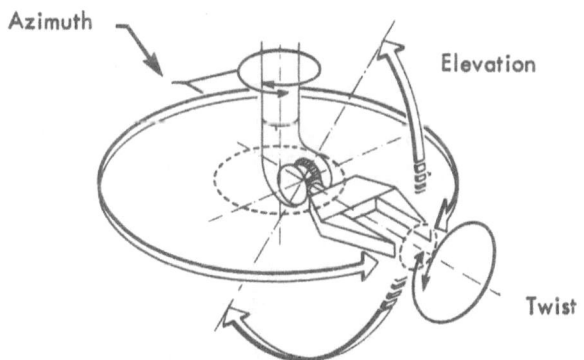

Fig. 3 Orientation motions.

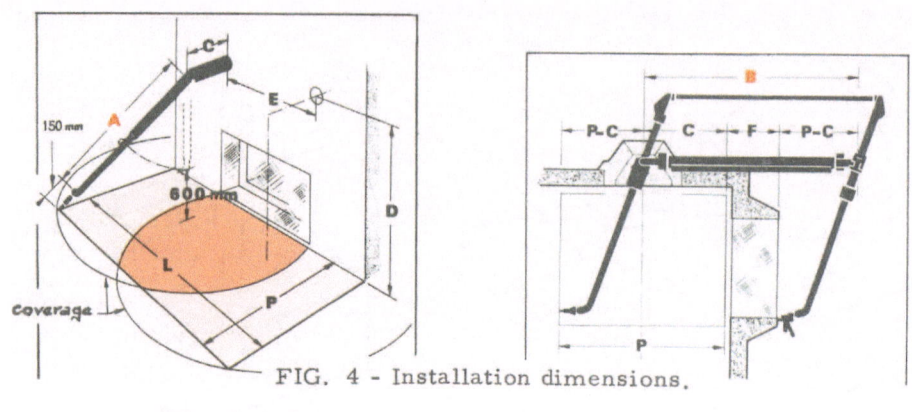

FIG. 4 - Installation dimensions.

1 manipulator Model G
2 manipulator Model 8

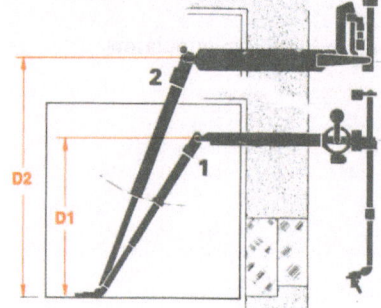

FIG. 5 - Required cell height for
same coverage with different
manipulator models.

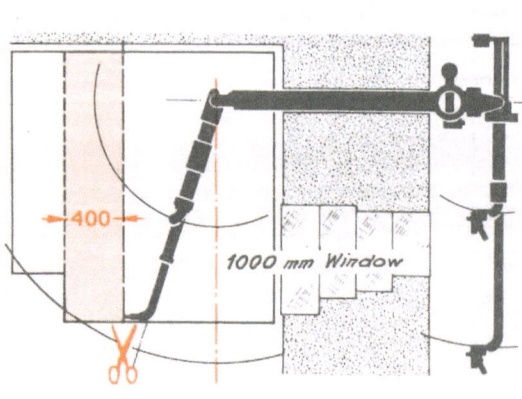

A. Telescopic Model

FIG. 6 - Mounting in limited depth.

B. Articulated Model Sized
For 100 mm Thick Lead
Shield

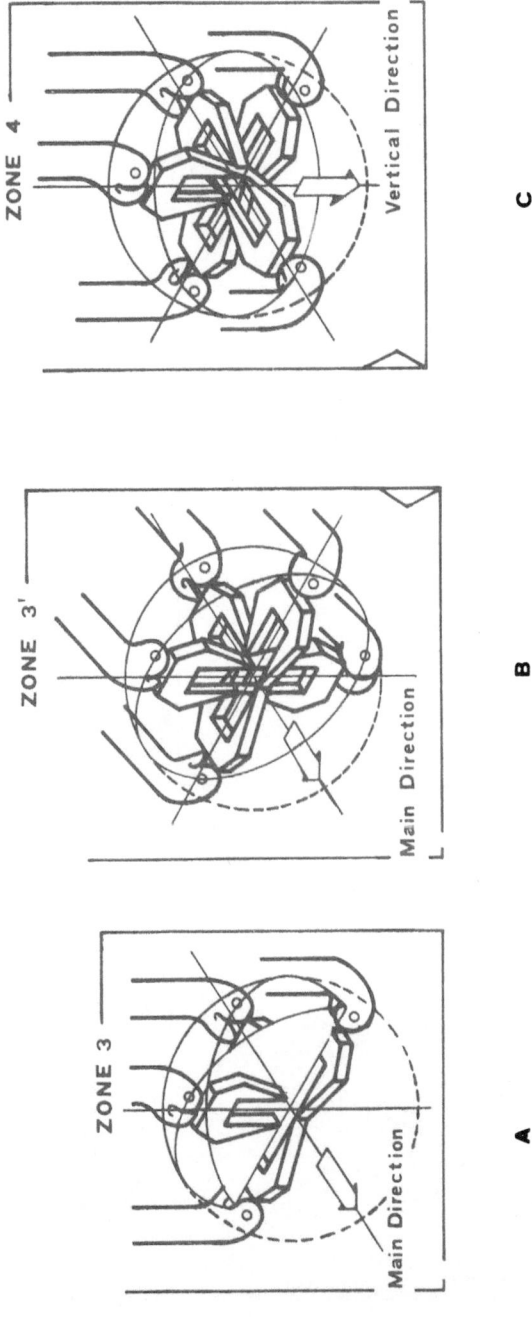

Fig. 7 Minimum pivoting possibilities in different zones. Note that main direction of zone 3 is similar to vertical motion in zone 4.

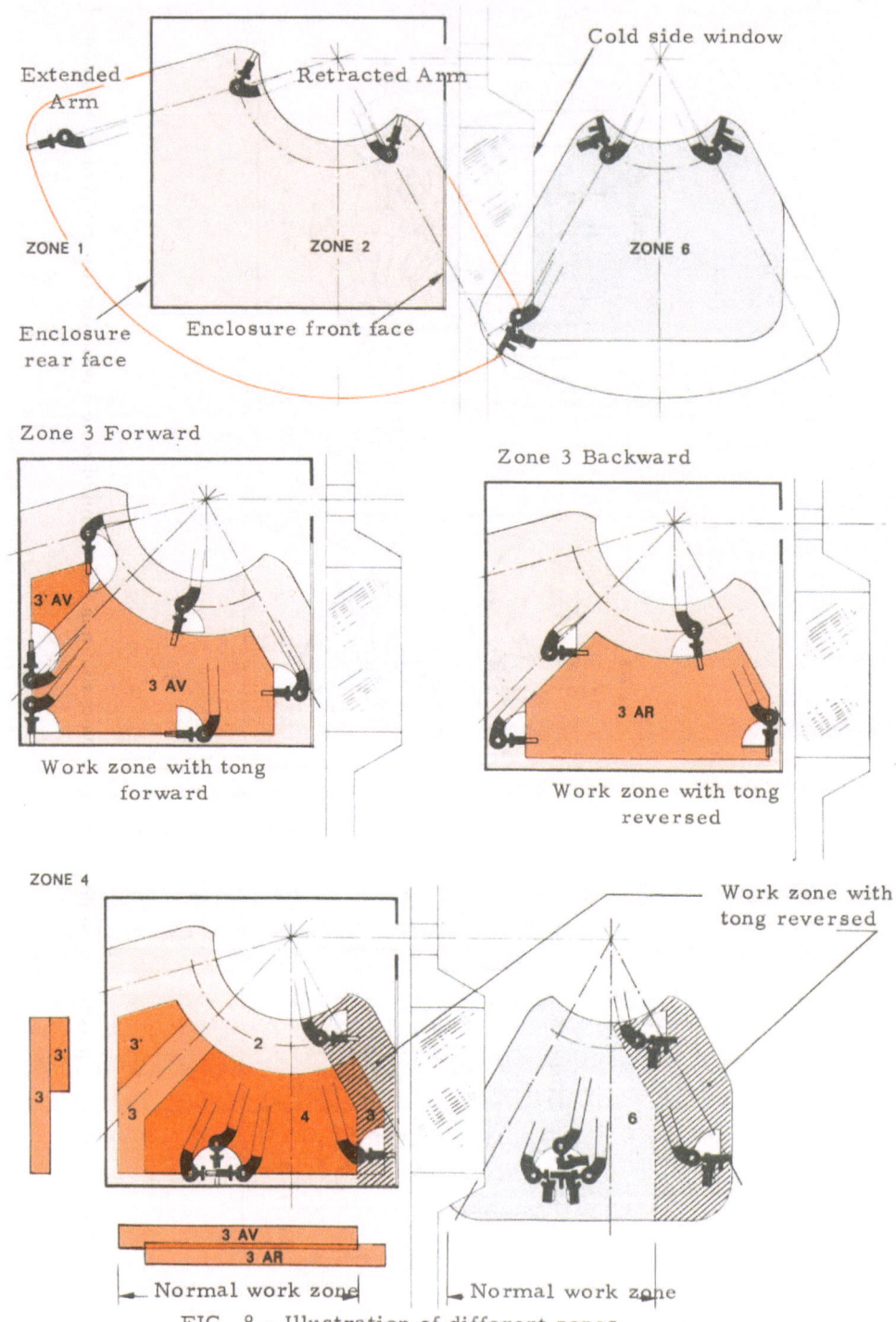

FIG. 8 - Illustration of different zones.

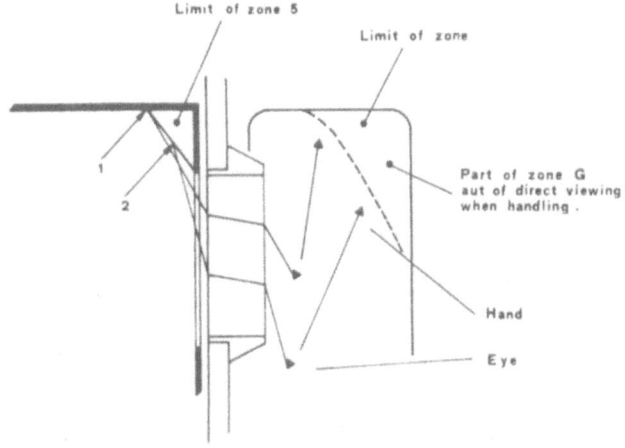

Fig. 9 Top view illustrating limits of zone 5 and 6.

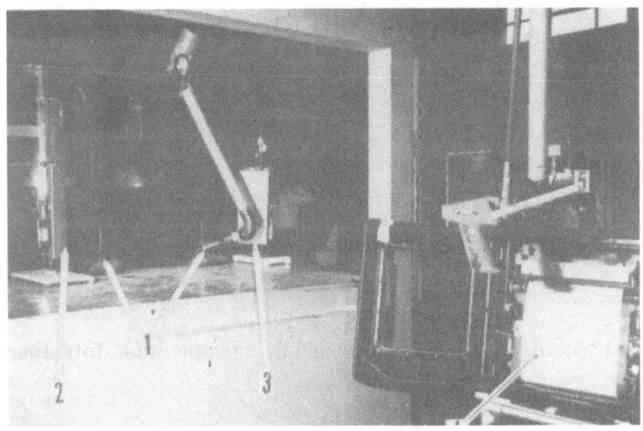

Fig. 10 Test procedure

1. Task test device.
2. Recorder "start" switch.
3. Recorder "stop" switch.
4. Recorder.
5. Manipulator "master" handle.

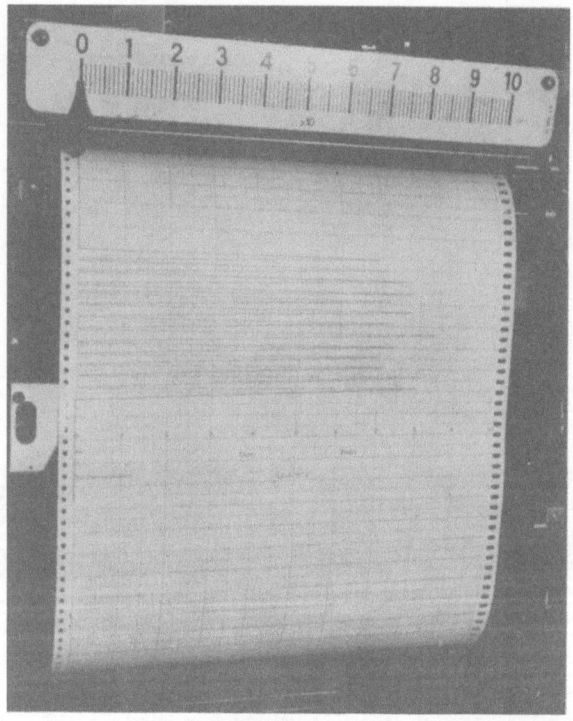

Fig. 11 Typical test record. Each pulse is recorded by one operation, total time is on the paper
length.

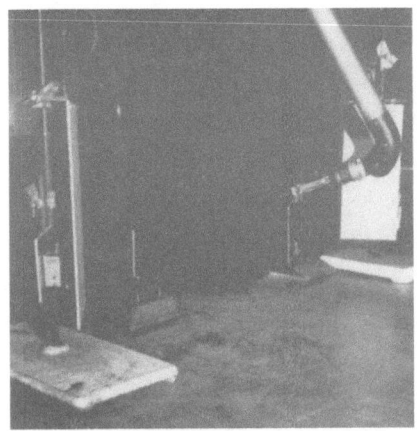

Fitting 0.5-kg block on pin after lifting off left pin.

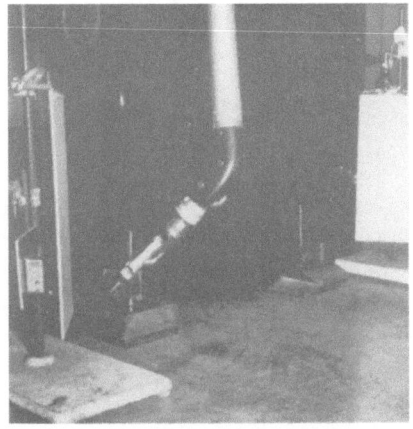

Removing 1 kg block from left side to fit
on right pin.

Fig. 12 A   Pick and place on pins task.

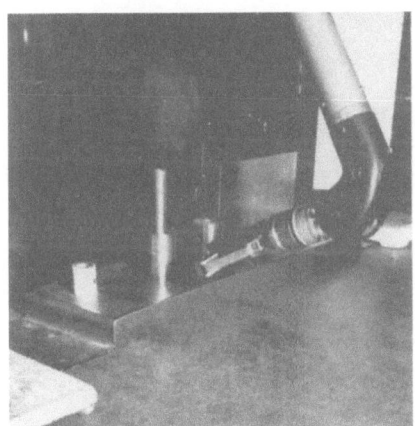

Start disassembling by removing
lock pin.

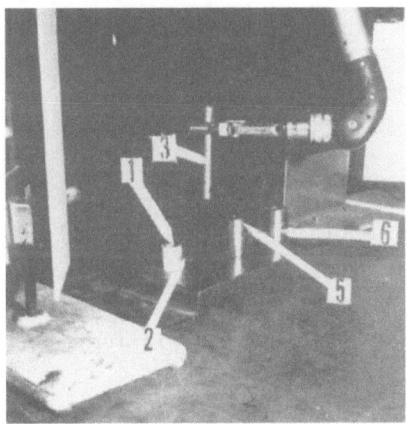

After fitting the pin on left block, remove
peg without moving 5; then 5 is removed.

Fig. 12 B   Assembling test

1. Large hole block
2. Locking pin
3. 1.27-cm peg

4. Fixed block
5. Movable block
6. Same as 4 and 5 but with a recess

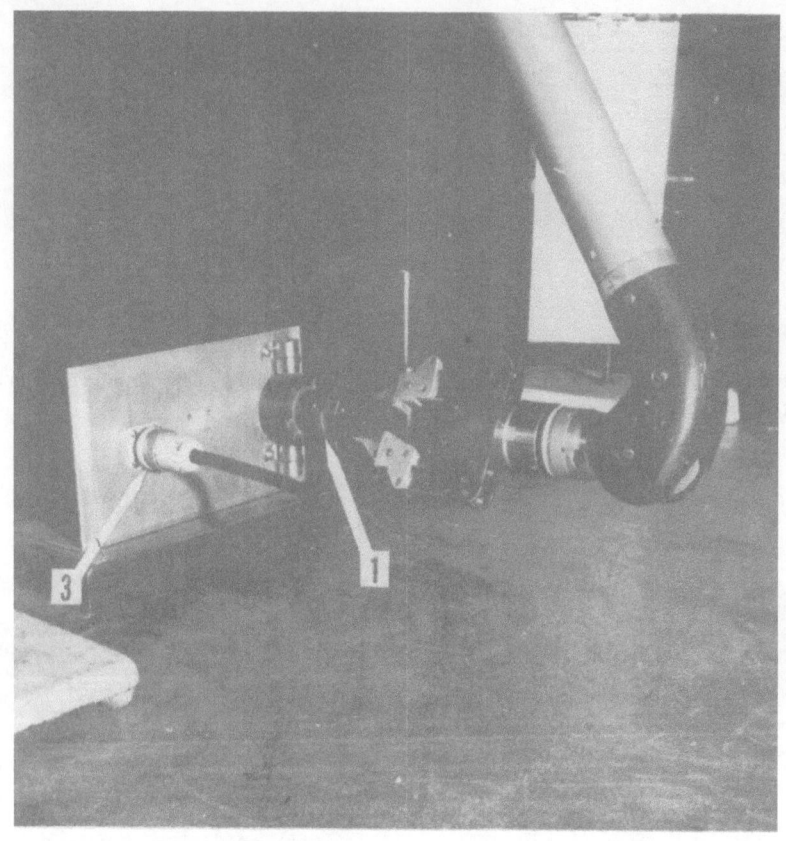

Fig. 13 Turning valve test and electric plug test.

1. Valve with flat knob
2. Electric plug
3. Rotary lock ring.

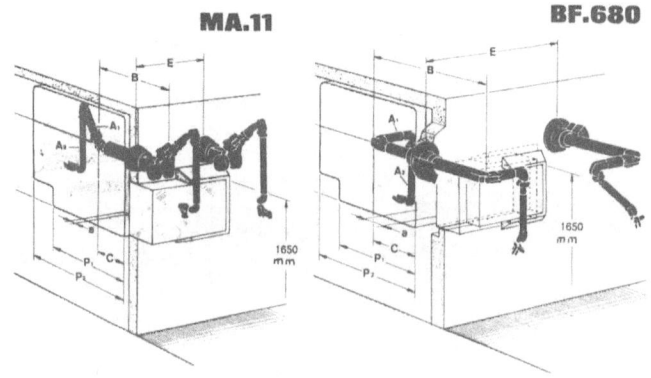

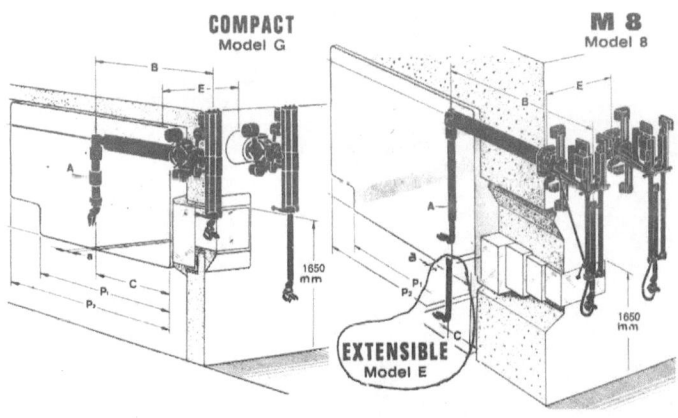

Fig. 14. Different master slave being tested.

B – Slave arm - 15 to 20 kg capacity.

Fig. 15   MA 22 Servo Bilateral Master Slave.
Prototype as working in June 1973.

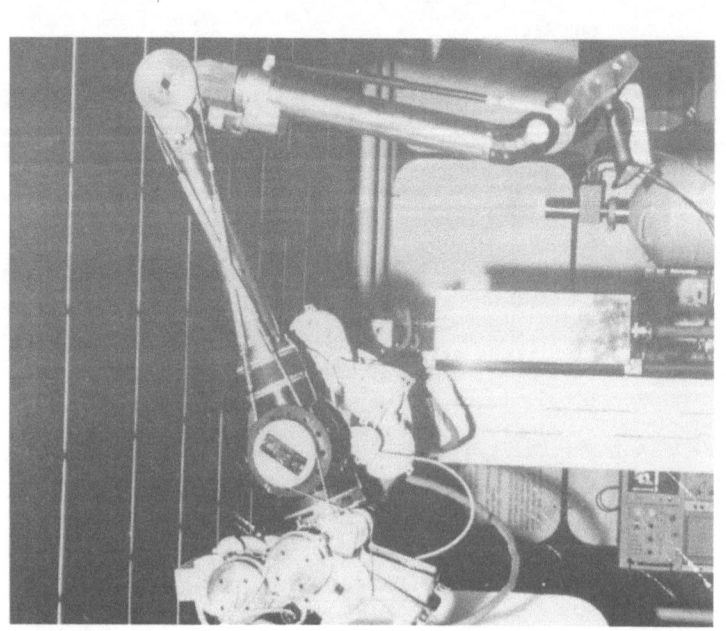

A – Master arm

Fig. 15   MA 22 Servo Bilateral Master Slave.
Prototype as working in June 1973.

T A B L E I
Coverage Analysis

Any object accessible to the tong can be twisted 360°
Possible use of the two other degrees-of-freedom depend on zones

Zone	Definitions	Limits
1	Maximum volume being covered at tong tips	Only limited by the geometry of the manipulator at limits of zone 1, tong can only twist
	Zones 2 to 6 are determined once the manipulator is fitted into a cell-limits are then due to cell walls and/or to touching cold cell face with the handle	
2	Maximum volume covered at tong tips inside the cell	Part of zone 1 inside the cell limited by cold-side interference
3	Volume where tong can pan at least 180 deg and tilt from horizontal to vertical down	Zone 3 can be limited by the main direction of the 180 deg pan motion-or considered with all main directions
3'	Volume where tong can pan at least 180 deg and tilt from vertical up to vertical down	Zone 3' is part of zone 3
4	Volume where tong can pan 360 deg and tilt from horizontal to vertical down	Zone 4 is the inside envelope of zone 3 (i.e., common part of zone 3 FWD and 3 BWD, etc.)
5	Volume covered by tong tips while viewing through the window	This zone is limited when the hand can reach a point where corresponding tong can be seen. Hand-to-eye distance is around 800 mm (32 in.)
6	Volume where the operator hand moves to cover zone 2	

T A B L E II

Time Efficiency of Manipulators Handlike Operations

OPERATIONS	HAND DATA	BILATERAL MASTER SLAVE (1) MA 11 PSM	(2) MA 11	(3) BF 680	(4) MODEL G PSM	(5) MODEL G	(6) MODEL 8 PSM	(7) MODEL 8	(8) HWM 100	(9) MODEL EHD	OPEN LOOP (10) SYNTELMANN	(11) PAR 3500	(12) MR 60
Task 1 – Pick and place on pins at 0.4 m 16 in	11.1 s	1.6	1.8	1.8	1.8	1.9	1.9	2.5	2	2.7	10	>50	>80
standard dev. σ%	6%			9%				7%		8%			
Lifting 0.5 Kg	0.2 s	3	4.5	4.5	4	4.5	5	7.5	6				
Lifting 1 Kg	0.3 s	3.5		4.5		4.5		7.5					
Fitting on pin and back 0.5 Kg	0.6* ⎰0.9	1.3	2	1.7	1.3	1.3	2.3	2.7	2.3				
1 Kg	1.1	1.3	2.5	2	1.6	2	2.4	3	2.4				
Average empty velocity 0.5 Kg	0.5 m/s	1.5	1.5	1.3	1.3	1.3	1.3	1.3	1.3				
in 0.4 m	same	1.5	1.5	1.4	1.6	1.5	1.3		1.4				
(16 in.) 1 Kg	20"/s	1.6	1.6	1.5	1.8	1.6	1.6	1.6	1.4				
Task 2 – Assembling	6.1 s	2	2.2	2.4				2.9		3.7	20	100	>100
Disassembling	5.1 s	1.5	2	2.2				2.7		3.2			
Standard dev. σ%				10%				10%		10%			

* Difference between two loads looks alike so factors are taken from average hand time.

(1) MA 11 French model by La Calhène also made as Model H by CRL
   PSM a new handle by La Calhène (see Figs.1 and 2)
(2) MA 11 with CRL regular handle
(3) French articulated arm by Sorige
(4) PSM mounted on CRL model G
(5) Regular hand on CRL model G
(6) PSM mounted on CRL model 8
(7) Regular mounted on CRL model 8
(8) German model with pistol handle
(9) CRL heavy-duty extended-reach with HD pistol handle
(10) German servo open-loop arm with exoskeleton control and limited force feedback
(11) PAR rate control
(12) Model by ACB (French) constant speed and on-off control.

T A B L E   III

Time Efficiency of Manipulators

	Hand Data	Time Efficiency							
		MA 11 or Model H PSM Handle	MA 11 or Model H	BF 680	Model G	Model 8	Model EHD*	Syntelmann*	P.
Task 3 = Valve									
Turning on	2.2 S	3.4	4	6		5.4	6	30	>5
Turning off	2.2 S	4	6	7		6	6.5		
Standard dev.(σ)	17%	8%	10%	13%		12%	7%		
Task 4 = Plug									
Flugging in	2.7 S	3.5	5.5	5.5	5.5	5	3.5	30	>5
Plugging out	1.4 S	6	6	4.5	6	7	6		
Standard dev.(σ)	15%	27%	25%	15%		22%		30%	
Strategy slowing factor		2.6	3	3.2	3	2.3	2	3	

Improved by high grasping force with a continuous twist.

T A B L E   IV

Time Efficiency Proposed General Chart

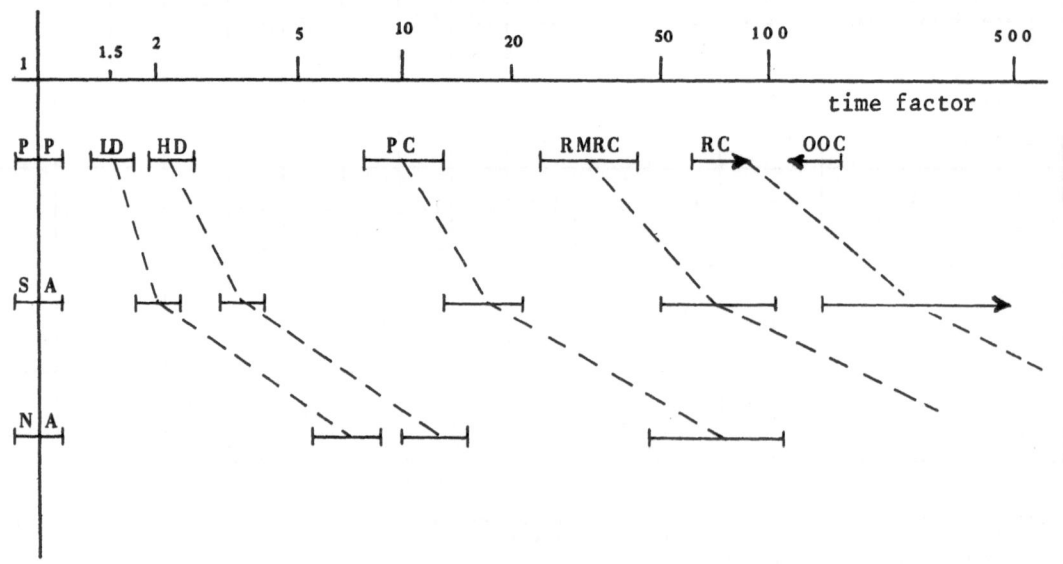

PP	pick and place	LD	light duty master slave
SA	simple assembly	HD	heavy duty bilateral master sla
NA	normal assembly	PC	position control-unilateral oft
RMRC	resolved motion rate control (MIT)		called master slave
RC	rate control	OOC	on-off control

CONTENTS

1. **Walking Machines**
   A.P. BESSONOV - N.V. UMNOV
   "The analysis of gaits in six-legged vehicles according to
   their static stabillty"

   I. KATO and Group of Bio-Engineering (Waseda
   University)
   "Information-Power machine with senses and limbs
   WABOT 1"

   R.B. McGHEE - D.E. ORIN
   "An interactive computer-control system for a quadruped
   robot"

   T. YAMASHITA - H. YAMADA
   "A study on stability of bipedal locomotion"

2. **Kinematics and Dynamics**
   I.I. ARTOBOLEVSKII - A.G. OVAKIMOV
   "A generalized method for solving a group of problems
   referring to the dynamics of manipulators with
   an electromechanical servo drive"

   M.S. KONSTANTINOV
   "A kinematical algorithm and dynamical point mass
   simulation applied in robots and manipulators"